W. Funk, V. Dammann, G. Donnevert
**Quality Assurance
in Analytical Chemistry**

1807–2007 Knowledge for Generations

Each generation has its unique needs and aspirations. When Charles Wiley first opened his small printing shop in lower Manhattan in 1807, it was a generation of boundless potential searching for an identity. And we were there, helping to define a new American literary tradition. Over half a century later, in the midst of the Second Industrial Revolution, it was a generation focused on building the future. Once again, we were there, supplying the critical scientific, technical, and engineering knowledge that helped frame the world. Throughout the 20th Century, and into the new millennium, nations began to reach out beyond their own borders and a new international community was born. Wiley was there, expanding its operations around the world to enable a global exchange of ideas, opinions, and know-how.

For 200 years, Wiley has been an integral part of each generation s journey, enabling the flow of information and understanding necessary to meet their needs and fulfill their aspirations. Today, bold new technologies are changing the way we live and learn. Wiley will be there, providing you the must-have knowledge you need to imagine new worlds, new possibilities, and new opportunities.

Generations come and go, but you can always count on Wiley to provide you the knowledge you need, when and where you need it!

William J. Pesce
President and Chief Executive Officer

Peter Booth Wiley
Chairman of the Board

Werner Funk, Vera Dammann, Gerhild Donnevert

Quality Assurance in Analytical Chemistry

Applications in Environmental, Food,
and Materials Analysis, Biotechnology,
and Medical Engineering

WILEY-VCH Verlag GmbH & Co. KGaA

The Authors

Werner Funk †

Dipl.-Ing. Vera Dammann
University of Applied Sciences
Gießen-Friedberg
FB KMUB
Wiesenstrasse 14
35390 Gießen
Germany

Dipl.-Ing. Gerhild Donnevert
University of Applied Sciences
Gießen-Friedberg
FB MNI
Wiesenstrasse 14
35390 Gießen
Germany

■ All books published by Wiley-VCH are carefully produced. Nevertheless, authors, editors, and publisher do not warrant the information contained in these books, including this book, to be free of errors. Readers are advised to keep in mind that statements, data, illustrations, procedural details or other items may inadvertently be inaccurate.

Library of Congress Card No.: applied for

British Library Cataloguing-in-Publication Data
A catalogue record for this book is available from the British Library

Bibliographic information published by the Deutsche Nationalbibliothek
The Deutsche Nationalbibliothek lists this publication in the Deutsche Nationalbibliografie; detailed bibliographic data are available in the Internet at http://dnb.d-nb.de.

© 2007 WILEY-VCH Verlag GmbH & Co. KGaA, Weinheim, Germany

All rights reserved (including those of translation into other languages). No part of this book may be reproduced in any form – by photoprinting, microfilm, or any other means – nor transmitted or translated into a machine language without written permission from the publishers. Registered names, trademarks, etc. used in this book, even when not specifically marked as such, are not to be considered unprotected by law.

Printed in the Federal Republic of Germany
Printed on acid-free paper

Cover SCHULZ Grafik-Design, Fußgönheim
Typesetting ProSatz Unger, Weinheim
Printing betz-druck GmbH, Darmstadt
Bookbinding Litges & Dopf Buchbinderei GmbH, Heppenheim

ISBN 978-3-527-31114-9

*In Memoriam
Professor Werner Funk*

Contents

Preface *XIII*

List of Symbols *XVII*

0	**Introduction** *1*	
0.1	General Differentiation of Analytical Processes *3*	
0.2	Quality of Analytical Processes and Results *4*	
0.3	The System of Analytical Quality Assurance *4*	
0.4	The Four-Phase Model of Analytical Quality Assurance *6*	
1	**Phase I: Establishing a New Analytical Procedure** *9*	
1.1	Introduction *9*	
1.1.1	Objectives of Phase I *9*	
1.1.2	When Are Characteristic Data Obtained? *9*	
1.1.3	The Progression of Phase I *10*	
1.1.4	Results of Phase I; Statistical Data *14*	
1.2	Calibration of the Fundamental Analytical Procedure *15*	
1.2.1	Establishment of an Analytical Range *16*	
1.2.2	Preparation of Standard Samples *16*	
1.2.3	Determination of the Calibration Function and Process Data *17*	
1.2.3.1	Process Data for the Linear Calibration Function *18*	
1.2.3.2	Process Data for the Second-Order Calibration Function *19*	
1.2.3.3	Calculating Analytical Results with the Aid of the Calibration Function *21*	
1.2.4	Verification of the Fundamental Calibration *23*	
1.2.4.1	Verification of Linearity *23*	
1.2.4.2	Verification of Precision *25*	
1.3	Analyses at Very Low Concentrations *29*	
1.3.1	Decision Limit *32*	
1.3.2	Determining the Minimum Detectable Value *34*	
1.3.2.1	Minimum Detectable Value, Determined Using the Distribution of Blank Values *34*	

Quality Assurance in Analytical Chemistry: Applications in Environmental, Food, and Materials Analysis, Biotechnology, and Medical Engineering, Second Edition. W. Funk, V. Dammann, G. Donnevert
Copyright © 2007 WILEY-VCH Verlag GmbH & Co. KGaA, Weinheim
ISBN: 978-3-527-31114-9

1.3.2.2	Minimum Detectable Value, Obtained Using the Calibration Function	35
1.3.3	Limit of Quantification	35
1.3.4	Quick Estimation	36
1.3.5	Estimation of the Decision Limit and Limit of Quantification Using the S/N Ratio	37
1.4	Validation of Individual Process Steps and Examination of Matrix Influences	37
1.4.1	Systematic Errors	37
1.4.1.1	Constant Systematic Errors, Additive Deviations	37
1.4.1.2	Proportional Systematic Errors, Multiplicative Deviations	38
1.4.2	Establishment and Assessment of the Recovery Function	38
1.4.2.1	Prerequisites for the Interpretation of the Recovery Function	39
1.4.2.2	Testing for Systematic Errors	40
1.4.3	Application of the Recovery Function	41
1.4.3.1	Checking Individual Process Steps	41
1.4.3.2	Determination of the Recovery Function to Prove the Influence of a Matrix	45
1.5	Additional Statistical Methods	46
1.6	Use of Internal Standards	46
1.6.1	Definition, Purpose	46
1.6.2	Conditions and Limitations of the Use of Internal Standards	47
1.6.3	Procedure	47
1.7	Preparing for Routine Analysis	49
1.7.1	Examination of the Time Dependency of Measured Values	49
1.7.1.1	Comparison of the "Within Batch" Standard Deviation (s_w) with the "Between Batches" Standard Deviation (s_b)	49
1.7.1.2	Determining the Need for Daily Adjustment of Analytical Equipment	51
1.7.1.3	The Trend Test	51
1.8	Summary of the Results of Phase I (Process Development): Documentation	54
2	**Phase II: An Analytical Process Becomes Routine; Preparative Quality Assurance**	**57**
2.1	Introduction	57
2.1.1	Objectives of Phase II	57
2.1.2	Execution of Phase II	57
2.1.3	Progression of Phase II	57
2.1.4	Results of Phase II	58
2.2	Selection of the Analytical Procedure	59
2.2.1	Specificity of the Procedure	60
2.2.2	Selectivity of the Analytical Procedure	60
2.2.3	Working Range	60
2.2.4	Calibration Function, Sensitivity, and Precision of the Procedure	60

2.2.5	Minimum Detectable Value and Limit of Quantification	*61*
2.2.6	Risk of Systematic Error	*61*
2.2.7	Effort, Costs	*61*
2.3	The "Training" Phase of the Process	*62*
2.4	Establishment of Quality Objectives to be Adhered to in Routine Usage	*64*
2.4.1	External Quality Requirements	*65*
2.4.2	Internal Quality Requirements	*66*
2.5	Control Samples for Internal Quality Assurance	*66*
2.5.1	Requirements of Control Samples	*66*
2.5.2	Types of Control Samples	*67*
2.5.2.1	Standard Solutions	*67*
2.5.2.2	Blank Samples	*67*
2.5.2.3	Natural Samples	*67*
2.5.2.4	Spiked Natural Samples	*68*
2.5.2.5	Synthetic Samples	*68*
2.5.2.6	Certified Reference Materials (CRMs)	*68*
2.5.3	Requirements for Producers of Control Materials	*69*
2.5.4	Applicability of Control Sample Types	*69*
2.6	The Control Chart System	*70*
2.6.1	Introduction: History of the Control Chart	*70*
2.6.2	Principle of a Control Chart	*72*
2.6.3	Average Run Length (ARL) and Evaluation of Control Charts	*73*
2.6.4	Derivation of the Average Run Length (ARL)	*74*
2.6.4.1	Examples of Theoretical Calculations	*75*
2.6.4.2	Analytical Example	*76*
2.6.5	Concept for the Preparation of Routine Quality Control	*78*
2.6.6	Evaluation of the Preliminary Period	*80*
2.6.6.1	Variance Analysis	*80*
2.6.6.2	Adherence to Required Quality Objectives	*80*
2.6.7	Types of Control Charts and Their Applications	*80*
2.6.7.1	Shewhart Charts	*81*
2.6.7.2	R-Chart (Range Control Chart)	*89*
2.6.7.3	Difference Chart	*96*
2.6.7.4	Standard Deviation Chart (s-Chart)	*98*
2.6.7.5	Target Value Charts	*99*
2.6.7.6	Cusum Chart	*100*
2.6.8	Summary of the Characterization of Control Charts	*112*
3	**Phase III: Routine Quality Assurance**	*115*
3.1	Introduction	*115*
3.1.1	Setting the Objectives of Phase III	*115*
3.1.2	Execution of Phase III	*115*
3.1.3	Progression of Phase III	*115*
3.2	Fundamental Measures of Internal Quality Assurance	*118*

3.2.1	The Laboratory and Laboratory Management 118
3.2.2	Personnel 119
3.2.3	Outfitting and Equipment 119
3.2.3.1	Performance Monitoring, Calibration and Adjustment of Measuring Equipment 119
3.2.3.2	Maintenance of Equipment 120
3.2.4	Materials 121
3.2.4.1	Certifying Sample Quality 121
3.2.4.2	Analysis-Related Materials 121
3.2.4.3	Control Samples for Routine Quality Control 122
3.2.5	Instituted Analytical Processes 122
3.2.6	Testing the Equivalency of Analytical Results 122
3.2.6.1	Testing the Equivalency for a Single Matrix 123
3.2.6.2	Testing the Equivalency in Different Matrices 126
3.2.7	Uncertainty of Measurements 130
3.2.7.1	New Terms According to the EURACHEM Guide 131
3.2.7.2	Overview of Common Procedures for the Determination of Measurement Uncertainty 133
3.2.7.3	Indication of Measurement Uncertainty in Test Reports 142
3.2.7.4	Interpretation of Measurement Uncertainty in the Context of Limit Value Monitoring 143
3.2.7.5	Summary 144
3.2.8	Reporting Analytical Results 145
3.3	Routine Quality Control 145
3.3.1	Trueness Control 146
3.3.1.1	General 146
3.3.1.2	Blank Value Monitoring 146
3.3.1.3	$\bar{x}$-Chart 148
3.3.1.4	Recovery Rate Control Chart 148
3.3.2	Precision Control 148
3.3.2.1	General 148
3.3.2.2	Precision Control Using an R-Chart 149
3.3.2.3	Securing Precision Using a Standard Deviation Control Chart 149
3.3.3	Revision of Quality Control Charts 150
3.3.4	Quality Assurance in the Case of Time-Consuming or Infrequent Analyses 150
3.4	Special Quality Problems in Routine Analysis 151
3.4.1	Matrix Effects 151
3.5	Corrective Measures 154
3.5.1	Sources of Error in Analytical Laboratories 154
3.5.2	Systematic Troubleshooting 155
3.5.2.1	Analytical Errors That Can Be Detected Using Statistical Quality Control Methods 161
3.5.2.2	Plausibility Checks 162
3.6	Documentation and Archiving 166

4	**Phase IV: External Analytical Quality Assurance** 169
4.1	Introduction 169
4.2	Audits 169
4.3	Interlaboratory (or Round Robin) Tests 170
4.3.1	Interlaboratory Tests for Process Standardization 171
4.3.2	Interlaboratory Tests as Proof of Laboratory Performance 171
4.3.3	Other Interlaboratory Tests 172
4.3.4	Planning and Execution of Interlaboratory (or Round Robin) Tests 173
4.3.4.1	Quality Management System of the Provider of an Interlaboratory Test 173
4.3.4.2	Planning the Interlaboratory Test 174
4.3.4.3	Interlaboratory Test Samples 175
4.3.5	Procedures for the Execution and Evaluation of Interlaboratory Tests 176
4.3.5.1	Interlaboratory Test Programs According to ISO 5725-2 177
4.3.5.2	The Youden Method of Interlaboratory Tests 179
4.3.5.3	Interlaboratory Tests According to ISO Guide 43 188
4.4	Effects of Internal Quality Assurance on the Results of Interlaboratory Tests 191
4 5	Conclusion 194

5	**Definitions** 195
5.1	Quality and Quality Management 195
5.2	Analytical Terms 197
5.3	Analytical Results 201
5.4	Deviation, Uncertainty 202
5.5	Materials, Samples 205
5.6	Statistical Tests 206
6	**References**

Appendix 1

A1	Sample Calculations 219
A1.1	Fundamental Calibration 219
A1.2	Linearity Tests 221
A1.2.1	Visual Linearity Test 221
A1.2.2	Second-Order Calibration Function 222
A1.2.3	Linearity Test: Goodness-of-Fit Test 224
A1.2.4	Variance Homogeneity Test 226
A1.2.5	Outlier Tests for Linear Calibration 228
A1.2.6	Securing the Lower Range Limit 230
A1.2.7	Decision Limit, Minimum Detectable Value, and Limit of Quantification 231
A1.2.8	Recovery Function 236

A1.2.9	Testing Analytical Results for Temporal Stability 239
A1.2.10	Trend Test 242
A1.2.11	Practice Phase: Checking the Analysis Quality Achieved Based on the Process Standard Deviation 243
A1.3	Phases II and III: Control Charts 244
A1.3.1	Blank Value Control Chart 245
A1.3.2	$\bar{x}$-Chart for Standard Solutions 246
A1.3.3	Recovery Rate (RR) Control Chart 247
A1.3.4	Verifying Precision by Means of R-Charts and s-Charts 249
A1.3.5	Testing for Series-Internal Drift 251
A1.3.6	RR-Control Chart by Addition of a Standard 253
A1.3.7	Cusum Chart 254
A1.3.8	Equivalency 258
A1.3.9	Standard Addition 261

Appendix 2

A2	Statistical Tables 263
A2.1	t-Table 263
A2.2	F-Table (95%) 264
A2.2	F-Table (99%) 265
A2.3	Grubbs Table 267
A2.4	χ^2-Table 268

Appendix 3

A3	Contents of the CD 269
A3.1	Checklists 269
A3.2	Instructions for Using the Calculation Examples 269
A3.3	Statistical Table Values 270

Subject Index 271

Preface to the Second Edition

In the more than 10 years since the appearance of the first edition of this book, analytical chemistry has not only undergone enormous scientific development due to technological progress but has witnessed a broadening in its range of applications, and has in part become a service that extends beyond its traditional boundaries. In parallel, awareness of the quality of analytical results has also increased. First, existing quality assurance systems were certified in some areas of analysis on the basis of the international quality management standards of the ISO 9000 series. These standards were originally created to increase the quality in goods production. Certification ensures that a quality assurance system exists and that tests are carried out in compliance with that system. However, an evaluation of the technical competence for the execution of specific tests did not take place thereafter. The special requirements for the technical competence of testing laboratories were accommodated in the mid-1990s through the introduction of accreditation in the quality standards based on the ISO Guide 25 "General Requirements for the Technical Competence of Calibration and Testing Laboratories" – implemented in Europe in 1994 through the standard series EN 45001 ff. This accreditation signified the confirmation not only of the conformity to particular technical rules, but also the efficacy of the laboratory by an impartial third party. In the meantime, the standard EN 45001 has been replaced by the ISO/IEC standard 17025 "General Requirements for the Technical Competence of Calibration and Testing Laboratories", which is recognized worldwide.

The procedures of analytical quality assurance described in the first edition of this book have worked satisfactorily in practice, and have now been further developed and amended. For the indication of an analytical result, the earlier (pre-1990) and mostly unfamiliar term "confidence interval" has been extended in content and renamed "measurement uncertainty". Triggered by the rising costs of the service of "analysis", the "equivalency" of faster, automated, more economical analytical procedures in comparison with reference procedures has taken on greater meaning.

This second edition accommodates all of these new requirements of quality assurance. To provide concrete assistance during day-to-day work in the laboratory, checklists and the elaborated mathematical examples in the Appendix are also available as Excel tables on the attached CD.

Quality Assurance in Analytical Chemistry: Applications in Environmental, Food, and Materials Analysis, Biotechnology, and Medical Engineering, Second Edition. W. Funk, V. Dammann, G. Donnevert
Copyright © 2007 WILEY-VCH Verlag GmbH & Co. KGaA, Weinheim
ISBN: 978-3-527-31114-9

We would like to dedicate this second edition to Professor Werner Funk, who sadly passed away much too early, in 1996.

Gießen, August 2006 *Vera Dammann*
  Gerhild Donnevert

Preface to the First Edition

When the German edition of "Quality Assurance in Analytical Chemistry" appeared in 1992, it was quickly accepted by analysts as a textbook on analytical quality assurance (AQA) and analytical quality control (AQC).

However, AQA and AQC have obtained higher importance since then. In recent years an increasing number of laboratories have applied for accreditation according to EN 45000 series. Quality management and quality assurance systems according to ISO 9000 ff are now certified in analytical laboratories – the "producers" of analytical results – too.

These latest trends were taken into consideration in this revised and updated English version. Some chapters were expanded and new ones were added. Moreover, some mistakes in the sample calculations (Appendix 1) were corrected.

The authors wish to thank the VCH Publishers for the support of this work and especially Mrs. Ann Gray for the competent translation of this book.

Gießen, June 1995
V. Dammann
G. Donnevert
W. Funk

List of Symbols

a	axis intercept of the linear calibration curve or the second-order calibration function
Δa	prediction interval of the axis intercept
a_c	axis intercept of the calibration curve for the fundamental analytical procedure
a_f	axis intercept of the recovery curve
ARL	average run length
b	slope of the linear calibration curve or the regression coefficient of the linear term of the second-order calibration function
b_c	slope of the calibration curve of the fundamental analytical procedure
b_f	slope of the recovery curve
b_s	slope of the spiking calibration function
c	regression coefficient of the quadratic term of the second-order calibration function
$CI(a_f)$	confidence interval of the axis intercept a_f
$CI(b_f)$	confidence interval of the slope b_f
$CI(\hat{x})$	confidence interval of the concentration $\hat{x}$
$CI(x_1)$	confidence interval at the lower end of the working range
$CI_{rel}(x_1)$	relative analysis precision at the lower end of the working range
$CI_{rel,req}$	required minimum precision of analytical results
$CI(x_i)$	confidence interval of the individual results
$CI(\bar{x})$	confidence interval of the mean values
$CI(\hat{y}_A)$	prediction interval of the regression line after outlier elimination for concentration x_A
$CI_x(y)$	confidence interval (in concentration units) of the observation y
d	leading distance of the V-mask on a cusum chart
d_i	residual or difference for a difference control chart or a Youden interlaboratory test program
$\bar{d}$	mean of the differences d_i
d_s	difference of the analytical results, $x_{As} - x_{Bs}$, of a suspected outlier laboratory (Youden interlaboratory test)
d_2	factor for calculating the standard deviation from the range

Quality Assurance in Analytical Chemistry: Applications in Environmental, Food, and Materials Analysis, Biotechnology, and Medical Engineering, Second Edition. W. Funk, V. Dammann, G. Donnevert
Copyright © 2007 WILEY-VCH Verlag GmbH & Co. KGaA, Weinheim
ISBN: 978-3-527-31114-9

Symbol	Description
D	deviation of the process mean from the reference value that should be detected
D_i	single difference between the result from the comparison procedure and the result of the reference procedure
D^*	suspected outlier single difference
$\bar{D}$	difference between the mean of the results from the comparison procedure and the results of the reference procedure
D_{Clo}, D_{Cup}, D_{Wlo}, D_{Wup}	factors for the calculation of the warning and control limits for range control charts
DS^2	difference of the sum of the squared deviations (variances)
e	distance of the new grand average from the old upper control limit of a control chart
e'	standardized distance e
$E(x)$	sensitivity
$E(\bar{x})$	sensitivity in the center of the range
f	degree of freedom
f_b	degree of freedom for the calculation of the variance s_b^2
f_i	degree of freedom for the calculation of the variance s_i^2
f_t	degree of freedom for the calculation of the variance s_t^2
f_w	degree of freedom for the calculation of the variance s_w^2
f_{CP}	degree of freedom for the data from the comparison procedure
f_R	degree of freedom for the data from the reference procedure
$F_{f1,f2;P}$	table value of the F-distribution
h	decision interval for the numerical evaluation of a cusum chart
i	running number of concentration steps or running number of subgroups/analytical series (1, 2, ... i, ... N)
j	running number of the analyses for each concentration step x_i or running number of analyses per subgroup/analysis series (1, 2, ... j, ... n_i)
k	factor for calculating the limit of quantification or the factor for calculating the expanded measurement uncertainty or the reference value for a cusum chart
K	decision limit for the numerical evaluation of a cusum chart or multiple of the standard deviation, s, by which the grand average of a control chart is shifted
K_{upper}	upper decision limit K
K_{lower}	lower decision limit K
l	number of laboratories participating in an interlaboratory (round robin) test program
L_0	ARL in the case of a process that is in control
L_1	ARL in the case of a process that is out of control
LCL	lower control limit for a control chart
LWL	lower warning limit for a control chart

List of Symbols

m_{exp}	assigned concentration of a parameter (expected value) in an accreditation interlaboratory test
n_i	number of analyses per concentration step, x_i, or per subgroup/analysis series
N	chosen number of concentration steps or number of subgroups or analysis series
N_a	number of multiple analyses of a sample
N_c	number of calibration concentration steps
N_f	number of concentration levels for the determination of the recovery curve
N_{A1}	number of concentration steps before elimination of an outlier
N_{A2}	number of concentration steps after elimination of an outlier
N_B	number of blank values
N_{CP}	number of analysis results for the comparison procedure
N_R	number of analysis results for the reference procedure
p	significance level; statistical probability
q	number of standard deviations for determining the scale of a cusum chart
Q_i	quotient of the result from the comparison procedure and the result of the reference procedure
Q^*	suspected outlier quotient
$\bar{Q}$	mean of the quotients Q_i
Q_{xx}, Q_{xy}, Q_{x^3}, Q_{x^4}, Q_{x^2y}	sum of the squares
r	correlation coefficient or repeatability in interlaboratory (round robin) tests
$r\,(P = 99\%, f)$	threshold value for testing the correlation coefficient
R	range
R_i	range of the series i
$\bar{R}$	mean of the ranges R_i
RR	recovery rate
RR_i	recovery rate of series i
$\overline{RR}$	mean recovery rate
s	standard deviation
s_{a_f}	standard deviation of the axis intercept a_f
s_{b_f}	standard deviation of the slope b_f
s_b	standard deviation between series (between batch)
s_d	standard deviation of the differences d_i or the average standard deviation of two analytical series in a mean value t-test or measurement accuracy in Youden interlaboratory (round robin) test programs

Symbol	Description
s_{exp}	spread for calculating the quality limits for proficiency tests
s_i	standard deviation of the measured values for analyses of standard samples with the concentration x_i
s_i^2	variance of the measured values for analyses of standard samples with the concentration x_i
s_r	repeatability standard deviation
s_t	total standard deviation
s_w	standard deviation within a series (within batch)
s_{xi}	single standard deviation in the estimation of the measurement uncertainty
s_{xo}	standard deviation of the method
s_{xo_1}	standard deviation of the method of the first analysis series
s_{xo_2}	standard deviation of the method of the second analysis series
s_{xo_c}	standard deviation of the method of the fundamental analytical procedure
s_y	residual standard deviation
s_{y_1}	residual standard deviation of the first-order calibration function or the first analysis series
s_{y_2}	residual standard deviation of the second-order calibration function or the second analysis series
$s_{y_{A1}}$	residual standard deviation of the calibration curve before eliminating outliers
$s_{y_{A2}}$	residual standard deviation of the calibration curve after eliminating outliers
s_{y_f}	residual standard deviation of the recovery curve
s_A	standard deviation of the analytical results of sample A (Youden interlaboratory test program)
s_B	standard deviation of the analytical results of sample B (Youden interlaboratory test program)
s_{Amax}	maximum acceptable value for the standard deviation s_A
s_{Bmax}	maximum acceptable value for the standard deviation s_B
s_{CP}	standard deviation of the results of a comparison procedure
s_D	standard deviation of the differences in joint t-tests
s_R	standard deviation of the results from the reference procedure or the reproducibility standard deviation from an interlaboratory test
s_{RC}	auxiliary value for the χ^2-test
s_{RR}	standard deviation of the recovery rate
s_1^2	variance of the measured values at the lower limit of the range
s_N^2	variance of the measured values at the upper limit of the range
S	cumulative sum (cusum)
S_N	cusum value after N series
S^-	negative systematic error on a cusum chart
S^+	positive systematic error on a cusum chart

$\left.\begin{array}{l}t_{f1,P};\ t_{f,P}\\ t_{f,\alpha};\ t_{f,\beta}\\ t_{f,(1-\alpha)}\end{array}\right\}$	table value of the t-distribution
T_i	sum of the value pairs x_{Ai}, x_{Bi} in a Youden interlaboratory (round robin) test program
$\bar{T}$	mean of the sums T_i
TV	test value for the performance of an F-test, t-test, or Grubbs test
$u(x_i)$	standard uncertainty
$u_c(y)$	combined uncertainty
$u(B)$	uncertainty on the basis of the laboratory components of the bias
$u(c_{CRM})$	uncertainty of the concentration of a certified reference material
$u(C_{ref})$	uncertainty components of the method bias
$u(e)$	uncertainty on the basis of random errors under repeatability conditions
$u(\text{bias})$	combined uncertainty from the method bias and laboratory components of the bias
$u(R_w)$	uncertainty components of the reproducibility within the laboratory
$u(\bar{R}_m)$	uncertainty of the mean recovery rate of the reference material
$U(y)$	expanded uncertainty
UCL	upper control limit of a control chart
UWL	upper warning limit of a control chart
V_s	spiking volume
V_{CP}	variation coefficient of the analytical results of the comparison procedure
V_R	variation coefficient of the analytical results of the reference procedure
V_{xo}	variation coefficient of the procedure
V_{xoR}	variation coefficient of the reference procedure
V_{xoCP}	variation coefficient of the comparison procedure
w	scale factor for cusum charts
x	concentration or analytical result
Δx	result uncertainty
Δx_{rel}	relative imprecision or result uncertainty
x_0	for the method of standard addition, the sample concentration estimated using the calibration function
x_1	concentration of the standard sample at the lower limit of the range
x_a	test value to secure the lower range limit
x_{act}	difference between the calculated concentrations of the spiked and original samples
x_{add}	added standard concentration or conventional true value in an interlaboratory test

x_c	calibration concentration
x_f	found concentration
x_i	concentration of the ith standard sample
x_{ic}	standard concentration no. i for the calibration of the fundamental analytical procedure
x_{if}	the found concentration calculated using the calibration function of the fundamental analytical procedure
x_{ij}	the jth analytical result for the subgroup/analysis series i
x_A	concentration of an eliminated outlier measurement
x_{CPi}	analytical result of the comparison procedure
x_{DL}	decision limit
x_{LQ}	limit of quantification
Δx_{LQ}	result uncertainty at the limit of quantification
x_{MDV}	minimum detectable value; capability of detection
x_N	concentration of the standard sample at the upper limit of the range
x_{Ri}	analytical result of the reference procedure
x_{S1}	first added spiking concentration
x_{S4}	maximum added spiking concentration
$\bar{x}$	mean of the standard concentrations x_i
$\bar{x}_A$	mean of the analytical results of sample A (Youden interlaboratory test)
$\bar{x}_B$	mean of the analytical results of sample B (Youden interlaboratory test)
x_{Ai}	analytical results of sample A (Youden interlaboratory test)
x_{Bi}	analytical results of sample B (Youden interlaboratory test)
$\bar{x}_i$	mean of the analytical results of a subgroup i
$\bar{x}_R$	mean of the analytical results of the reference procedure
$\bar{x}_{CP}$	mean of the analytical results of the comparison procedure
$\bar{\bar{x}}$	grand average
$\bar{x}_c$	mean of the calibration concentrations
$\hat{x}$	concentration of an analyzed sample with the measured value $\hat{y}$, as calculated using the calibration function
$\hat{x}_s$	concentration of a spiked sample calculated using the calibration function
x^2	test value for the χ^2-test
y_0	measured value of the bulk sample in the method of standard addition
$\hat{y}_0$	axis intercept of the spiking calibration function calculated using linear regression
y_A	eliminated outlier measurement
y_B	blank value
y_a	auxiliary value for the determination of x_a
y_c	critical value of the measured value (auxiliary value for the determination of x_{DL})

Symbol	Description
y_i	measured value of the ith standard sample
y_{ij}	jth measured value for the concentration x_i
$y_{S1\ldots S4}$	measured values of the spiked samples
$\bar{y}$	mean of the measured values y_i from the calibration experiment
$\bar{y}_B$	mean blank value
$\Delta\bar{y}_B$	prediction interval for future blank values
$\hat{y}$	measured value of an analyzed sample
$\hat{y}_i$	measured value of the standard concentration x_i calculated using the calibration function
$\hat{y}_A$	measured value of the standard concentration x_A calculated using the calibration function
Z	Z score
α	significance level
β	significance level
Δ^2	successive difference distribution
δ	method bias
η	recovery rate
Θ	angle of the V-mask
$\chi^2\left(n-1;\dfrac{\alpha}{2}\right)$, $\chi^2\left(n-1;1-\dfrac{\alpha}{2}\right)$	threshold values of the χ^2 (chi-squared) distribution for the s-control chart

0
Introduction

The results of physical and chemical measurements and analyses touch on practically every aspect of modern life: for example, we demand proof of unadulterated foodstuffs, we expect therapeutically effective doses of pharmaceuticals, and we feel threatened by environmental pollution when we cannot perceive pollutants directly with our own senses and instead picture in our imagination water, soil, and air quality based on a system of threshold values and analytical results. Even our own bodies become the object of analytical examination when, whether preventatively or in case of actual illness, we undergo clinical diagnosis. Analytical results are produced in both great number and at great cost, with varying objectives (Table 0-1).

Several subcategories of analytical science are regulated by law, which determines the type and frequency of the analyses performed. Examples of such areas are:

- medicine: epidemic control laws
- pharmacy: drug purity laws
- production of materials: chemical regulations
- health protection/occupational health and safety: hazardous materials, foodstuff laws
- environmental protection: drinking water purity, waste water purity, air emissions standards, etc.

The objective of analytical work is the achievement of reliable analytical results of a defined quality. Quality characteristics of analytical processes are therefore:

- *Specificity*: the ability of an analytical process to register the desired analytes in all relevant forms.

- *Selectivity*: the ability of an analytical process to register only the desired analyte, while other components in or characteristics of the sample – known as the matrix – do not influence the result.

- *Sensitivity*: the change in measured value per change in analyte concentration.

- *Accuracy*: in the sense of both *trueness* (lack of systematic errors) and *precision* (measurement of differences between results, as obtained by repeated use of an established analytical process on the same sample; imprecision is caused by random errors).

Quality Assurance in Analytical Chemistry: Applications in Environmental, Food, and Materials Analysis, Biotechnology, and Medical Engineering, Second Edition. W. Funk, V. Dammann, G. Donnevert
Copyright © 2007 WILEY-VCH Verlag GmbH & Co. KGaA, Weinheim
ISBN: 978-3-527-31114-9

Table 0-1 Purposes and uses of analyses.

Analytical purpose	Special requirements of the analytical process	Examples
Detection of harmful agents, screening	– high degree of specificity – low detection limit – rapid execution – economical	– preventive medical examinations (e.g., blood sugar test sticks) – toxicology and forensic medicine: identification of poisons and drugs – toxin identification and cause determinations in environmental incidents
Decision making	– high degree of specificity – high degree of accuracy	– proof in criminology – medical diagnostics – evaluation of toxic emissions from facilities (threshold limits) – determination of fees and penalties (e.g., waste water emissions, solid waste disposal, atmospheric emissions) – quality control testing of raw material upon receipt from suppliers
Ensuring the protection of people and the environment	– high degree of specificity – high degree of accuracy	– warning systems (e.g., gas detection systems) – foodstuff inspections – quality control of pharmaceuticals, chemicals, fuels, and other materials
Evaluation of performed measures	– high degree of specificity – high degree of accuracy	– analyses related to the licensing of new or modified technical installations – environmental clean-up – evaluation of medical therapies
Monitoring, process supervision, and process regulation	– high degree of specificity – high degree of accuracy – rapid execution	– process supervision of foodstuff, pharmaceutical, chemical, and material production – emissions monitoring of environmentally relevant facilities – progress control of long term medical therapies
General information	– high degree of specificity – high degree of accuracy – rapid execution	– basic research in all fields – preparation of environmental databases

The task of the analyst is to determine the quality of each analytical process he or she uses, and to improve, guarantee, and document it where necessary, so that the achieved quality is maintained at each point during routine analysis. Only then is an analytical process considered reliable.

The tasks required of analytical physics and chemistry may generally be seen as the determination of the answer to one of the following five basic questions:

1. Is a substance present in a given sample or not?
2. In which range of concentration is the substance in question present?
3. Is the concentration below appropriate threshold levels?
4. Has a threshold level been exceeded?
5. In what concentration (plus or minus statistical error) is the substance in question present in the sample?

These different questions necessitate varying requirements in the quality of the analytical procedure being used.

0.1
General Differentiation of Analytical Processes

Analytical processes may be categorized into five groups based on the objective, the effort required, and the required degree of information to be provided by the results.

1. *Standard methods*: which may be national standard procedures, or international (e.g., ASTM, ISO, ECCLES, CEN, IUPAC, etc.) guidelines and standards, or procedures described by legal regulations. Standard methods represent the accepted latest state of engineering; their use may be prescribed and/or obligatory within a framework of regulations and guidelines.
2. *Comparable laboratory procedures*: quantitative analytical processes which vary in procedure from standard methods, but the equivalence of which is verified using standard methods.
3. *Other quantitative analytical procedures* which have various uses and objectives in the laboratory outside of governmentally regulated areas. These also include the latest analytical procedures for which standards have not yet been established.
4. *Field methods*: analytical processes which can be performed outside the laboratory. With the aid of objective measurement methods, field methods can yield quantitative or semi-quantitative results depending on previous handling of the sample and quality assurance procedures.
5. *Orientation tests/Screening procedures* [88] assess subjectively; they can qualitatively prove the presence of a certain substance and allow a concentration estimate to be made which is accurate to within one order of magnitude.

Procedures 4 and 5 fall under the term "ready-to-use" procedures [121].

0.2
Quality of Analytical Processes and Results

The appropriate analytical process for obtaining the answer to a specific question needs not always be the scientifically "best" procedure (Table 0-2). However, it is important that the results of an analysis, obtained through adherence to legal regulations, should if necessary be defensible in court, or in other words legally binding.

Table 0-2 Precision requirements of analytical procedures as related to the analytical problem.

Question	Quality required of an analytical process
Is a substance present or not?	Yes/no decision; a sufficiently sensitive orientation test is applicable.
In which concentration range is the substance present?	An orientation test indicating a corresponding gradation of concentration ranges is applicable.
Does the concentration lie far below a verifiable threshold limit?	A highly precise analytical result is not necessary; orientation tests and field methods are applicable.
Has a threshold limit been exceeded?	A highly accurate (meaning both precision *and* trueness) analytical result is required, i.e., only reference procedures and, if need be, comparable laboratory procedures are applicable.
In what concentration, ± confidence interval, is the substance in question present in the sample?	A highly accurate (meaning both precision and trueness) analytical result is required. If the analytical result is legally binding, only standard methods or comparable laboratory procedures are acceptable. Otherwise, other analytical procedures of proven reliability are applicable.

0.3
The System of Analytical Quality Assurance

The objective of analytical quality assurance is to obtain reliable analytical results, the accuracy of which is determined, regularly verified, and documented.

Test series in the context of quality control of pharmaceutical products and toxic chemicals are above all subject to the rules of Good Laboratory Practice (GLP), which defines the requirements for the objectives, planning, performance, monitoring, recording, evaluation, and documentation of laboratory test series [19, 20, 181, 182, 184]. The OECD internationally organized GLP system [166–174] is *not* part of analytical quality assurance and is therefore not discussed in this book.

Table 0-3 Responsibilities and relationships of external analytical quality assurance.

Certification of the capability of laboratories to execute certain analyses on certain types of samples	External quality control	Monitoring of measuring and inspection equipment (instrument calibration)
National Accreditation Council: Responsibilities: international harmonization, coordination of national activities in the field of accreditation and recognition of testing laboratories, calibration laboratories, certification and inspection bodies; administration of national accreditation and recognition registers; definition of accreditation criteria, accreditation of accreditation agencies ↓		*Bureau International des Poids et Mesures (BIPM)/ International Bureau of Weights and Measures (Paris):* Creation of the experimental basis for the SI system; definition of the international standards; calibration of national reference measuring instruments ↓
Certification and accreditation bodies and agencies (e.g., inspection authorities): Certification and accreditation of laboratories		*National Metrology Institute (i.e., the Physical-Technical Federal Agency in Germany):* Preservation of the national standards; calibration of the reference measuring instruments of the calibration service ↓
		Calibration service: Calibration of measuring instruments: determination and documentation of measurement accuracy
	Reference laboratories: Execution of interlaboratory tests, distribution of certified reference material, assessment ↓	
	Laboratories, which carry out analyses	

All steps in the production and evaluation of analytical results in commercially and nationally regulated areas are bound in a hierarchically organized system of internal and external laboratory quality assurance (Table 0-3).

At national and European levels, the legislator defines limits, goals, and quality requirements; prescribes the inspection of product quality and, if necessary, also of reference procedures; stipulates the organization of an analytical quality assurance system; and also supports, in part, the founding and operation of certification, accreditation, and inspection bodies [21, 102, 144, 152, 177, 180, 185, 189, 207].

International (i.e., ISO, CEN) and national (i.e., DIN, BSI, AFNOR, NEN) standards define terms; specify quality assurance systems, courses of action, analytical procedures, and measuring instruments; and provide the basis for the evaluation of the operation and the competence of laboratories in the process of accreditation and certification.

Certification verifies that a laboratory has established and maintained a functional quality management system [66, 76].

For *accreditation*, a laboratory must prove that it can competently carry out certain analyses using certain types of samples and that it fulfills all corresponding regulations and standards [142].

Every chemical analysis is based in some way on the measurement of a physical or chemical quantity with the assistance of measuring instruments. Ensuring the accuracy of these measuring instruments is the objective of scheduled *instrument calibration*.

The inspection of the laboratory's competence also includes the participation in *interlaboratory tests* as a means of comparison with other laboratories. The organizing laboratory in the interlaboratory tests produces a large amount of a very homogeneous sample and sends smaller subsamples to the participating test laboratories for analysis. The results of the participants' analyses are then collected and evaluated. The assessment of a laboratory's performance is conducted on the basis of the trueness and precision of its results.

Interlaboratory tests are not only carried out as an assessment of a laboratory's analytical performance, but also:

- as a test of the practicality of new analytical procedures,
- to determine the characteristics of the procedures, and
- to identify the characteristics of the reference material.

0.4
The Four-Phase Model of Analytical Quality Assurance

The determination and, where necessary, the improvement and maintenance of the quality of analytical results are prerequisites for accrediting a laboratory for the use of analytical processes to analyze certain types of sample. Even when an accreditation is not intended, it is in one's own interest to carry out analytical quality assurance.

For the analyst, the quality assurance characteristics can be constructed more or less chronologically in four phases, as described below:

- **Phase I**: The quality characteristics of a new analytical process in need of calibration are determined and if necessary the process is improved and documented.

- **Phase II**: Preparative quality control. An analytical process for which the quality characteristics have already been documented is made operational for routine analysis. This includes achieving and maintaining sufficient analytical quality before routine analysis begins.

- **Phase III**: All internal laboratory quality assurance measures of routine analysis, accompanied by

- **Phase IV**: External laboratory quality assurance in the form of interlaboratory tests and external audits.

Only by tying together all analytical activities into a closed system of both internal and external laboratory quality assurance can the reliability of analytical results be guaranteed: these are, on the one hand, legally defensible and, on the other hand, they can be used to make comparisons of trend experiments, establish registers of analytical values, etc.

1
Phase I: Establishing a New Analytical Procedure

1.1
Introduction

1.1.1
Objectives of Phase I

Before a new analytical procedure, especially one requiring calibration, can be used for routine analysis, the individual steps need not only be determined but, where necessary, they must also be optimized, and the entire fundamental analytical procedure must be verified for its performance. The performance characteristics obtained in this way are documented and/or published with the description of the analysis and they form the basis for later quality assurance in routine analysis.

1.1.2
When Are Characteristic Data Obtained?

The statistical methods described below find their application primarily during the establishment of a new analytical process. In addition, they are suitable for providing the analyst with information about recent analytical quality achieved during the testing and training phase of an analytical process that is in need of calibration (see also Section 2.3).

The process data (linear calibration function including precision measures) should be determined anew with each new calibration of the analytical process, for example after:

- changing reagents (new batches),
- technical intervention in analytical equipment (after technical modifications, maintenance and repair, for example bulb changes in photometry),
- changes in staff,
- etc.

Quality Assurance in Analytical Chemistry: Applications in Environmental, Food, and Materials Analysis, Biotechnology, and Medical Engineering, Second Edition. W. Funk, V. Dammann, G. Donnevert
Copyright © 2007 WILEY-VCH Verlag GmbH & Co. KGaA, Weinheim
ISBN: 978-3-527-31114-9

1.1.3
The Progression of Phase I

Scheme 1-1 Procedure of Phase I: flowchart.

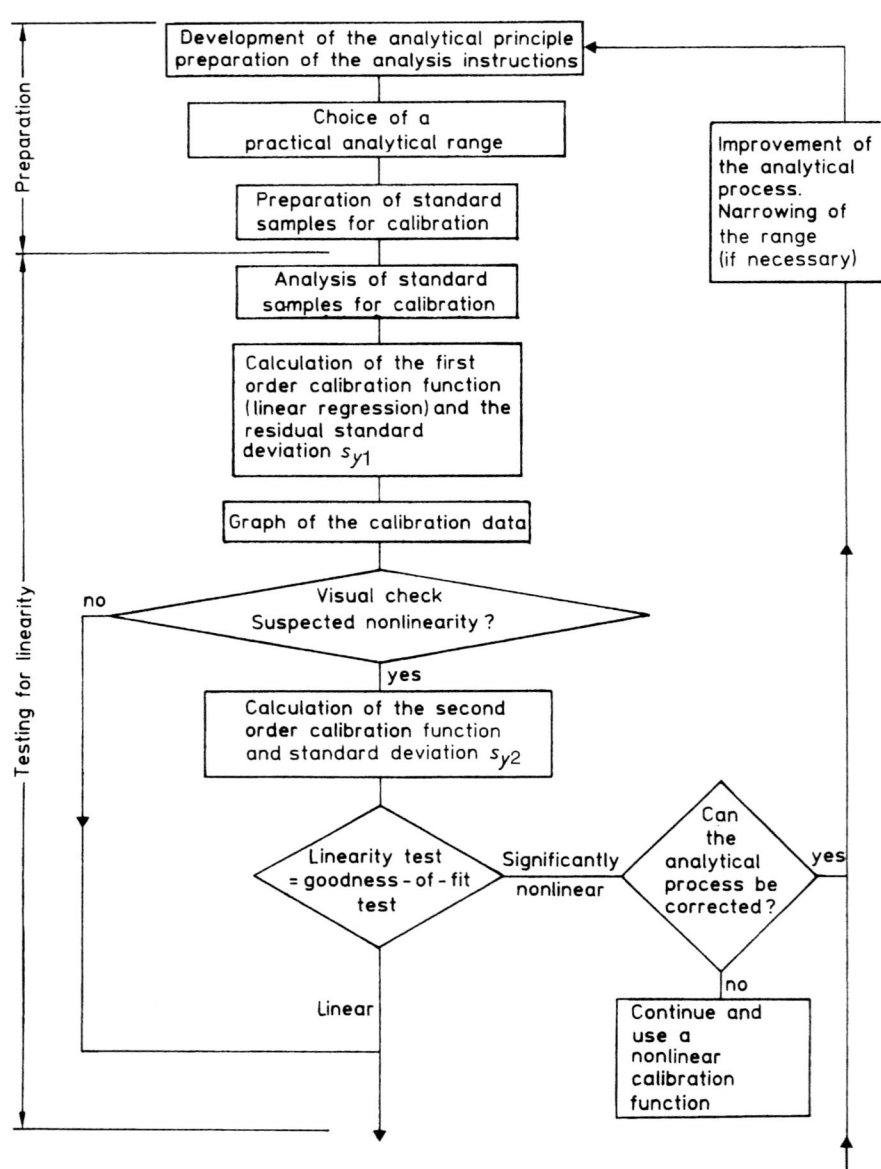

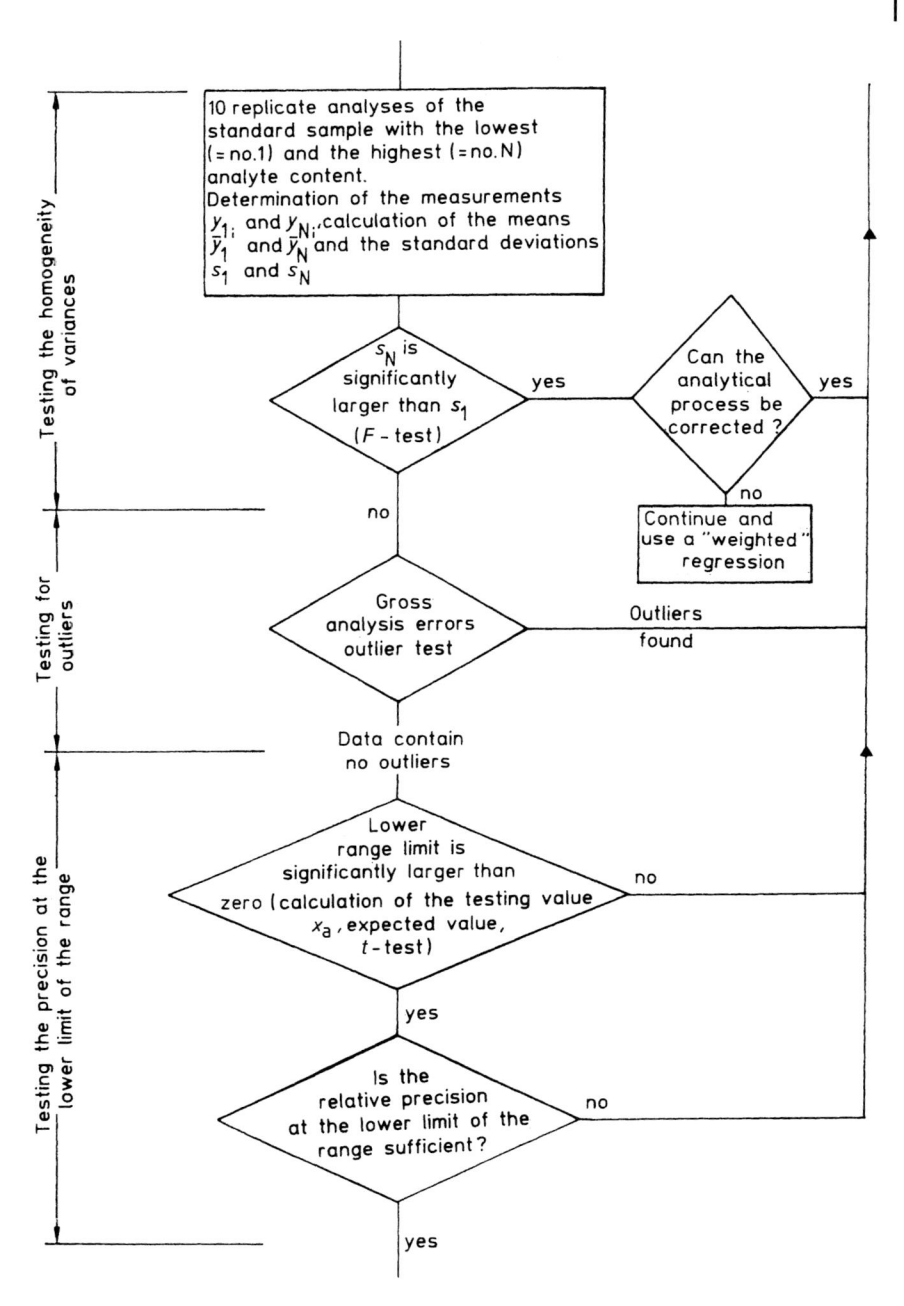

1.1 Introduction

```
                    │
                    ▼
         ┌──────────────────────────────┐
         │ 10 replicate analyses of the │
         │ standard sample with the lowest │
         │ (= no.1) and the highest (= no.N) │
         │ analyte content.             │
         │ Determination of the measurements │
         │ $y_{1i}$ and $y_{Ni}$; calculation of the means │
         │ $\bar{y}_1$ and $\bar{y}_N$ and the standard deviations │
         │ $s_1$ and $s_N$              │
         └──────────────────────────────┘
                    │
                    ▼
```

Testing the homogeneity of variances: s_N is significantly larger than s_1 (F-test) — **yes** → Can the analytical process be corrected? **yes** → (loop back). **no** → Continue and use a "weighted" regression.

Testing for outliers: Gross analysis errors outlier test → Outliers found → (loop back). Else → Data contain no outliers.

Testing the precision at the lower limit of the range: Lower range limit is significantly larger than zero (calculation of the testing value x_a, expected value, t-test) — **no** → (loop back). **yes** → Is the relative precision at the lower limit of the range sufficient? **no** → (loop back). **yes** → (continue).

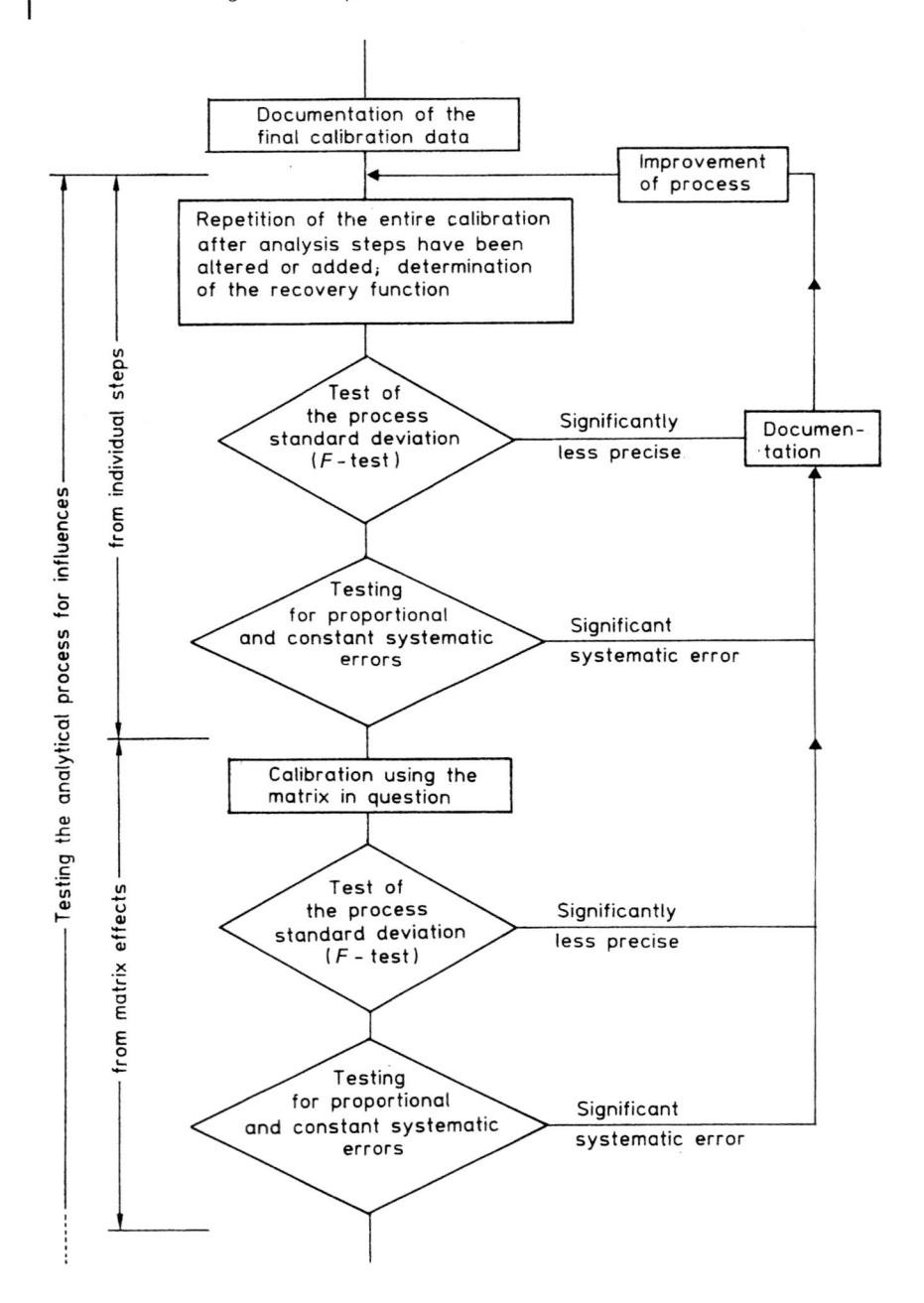

1 Phase I: Establishing a New Analytical Procedure

```
                    │
                    ▼
        ┌───────────────────────┐
        │ Documentation of the  │
        │ final calibration data│◄──────────┐  Improvement
        └───────────────────────┘           │  of process
                    ▲                       │
                    │                       │
        ┌───────────────────────────┐       │
        │ Repetition of the entire  │       │
        │ calibration after analysis│       │
        │ steps have been altered   │       │
        │ or added; determination   │       │
        │ of the recovery function  │       │
        └───────────────────────────┘       │
                    │                       │
                    ▼                       │
              ╱─────────────╲               │
             ╱   Test of     ╲  Significantly  ┌──────────┐
            ╱  the process    ╲─less precise──►│ Documen- │
            ╲ standard deviation╱              │  tation  │
             ╲   (F-test)     ╱                └──────────┘
              ╲─────────────╱                       ▲
                    │                               │
                    ▼                               │
              ╱─────────────╲                       │
             ╱   Testing     ╲                      │
            ╱ for proportional╲  Significant        │
            ╲and constant systematic─systematic error┤
             ╲    errors     ╱                      │
              ╲─────────────╱                       │
                    │                               │
                    ▼                               │
        ┌───────────────────────┐                   │
        │ Calibration using the │                   │
        │   matrix in question  │                   │
        └───────────────────────┘                   │
                    │                               │
                    ▼                               │
              ╱─────────────╲                       │
             ╱   Test of     ╲  Significantly       │
            ╱  the process    ╲─less precise────────┤
            ╲ standard deviation╱                   │
             ╲   (F-test)     ╱                     │
              ╲─────────────╱                       │
                    │                               │
                    ▼                               │
              ╱─────────────╲                       │
             ╱   Testing     ╲                      │
            ╱ for proportional╲  Significant        │
            ╲and constant systematic─systematic error┘
             ╲    errors     ╱
              ╲─────────────╱
                    │
                    ▼
```

←——— Testing the analytical process for influences ———→
←——— from individual steps ———→
←——— from matrix effects ———→

1.1 Introduction

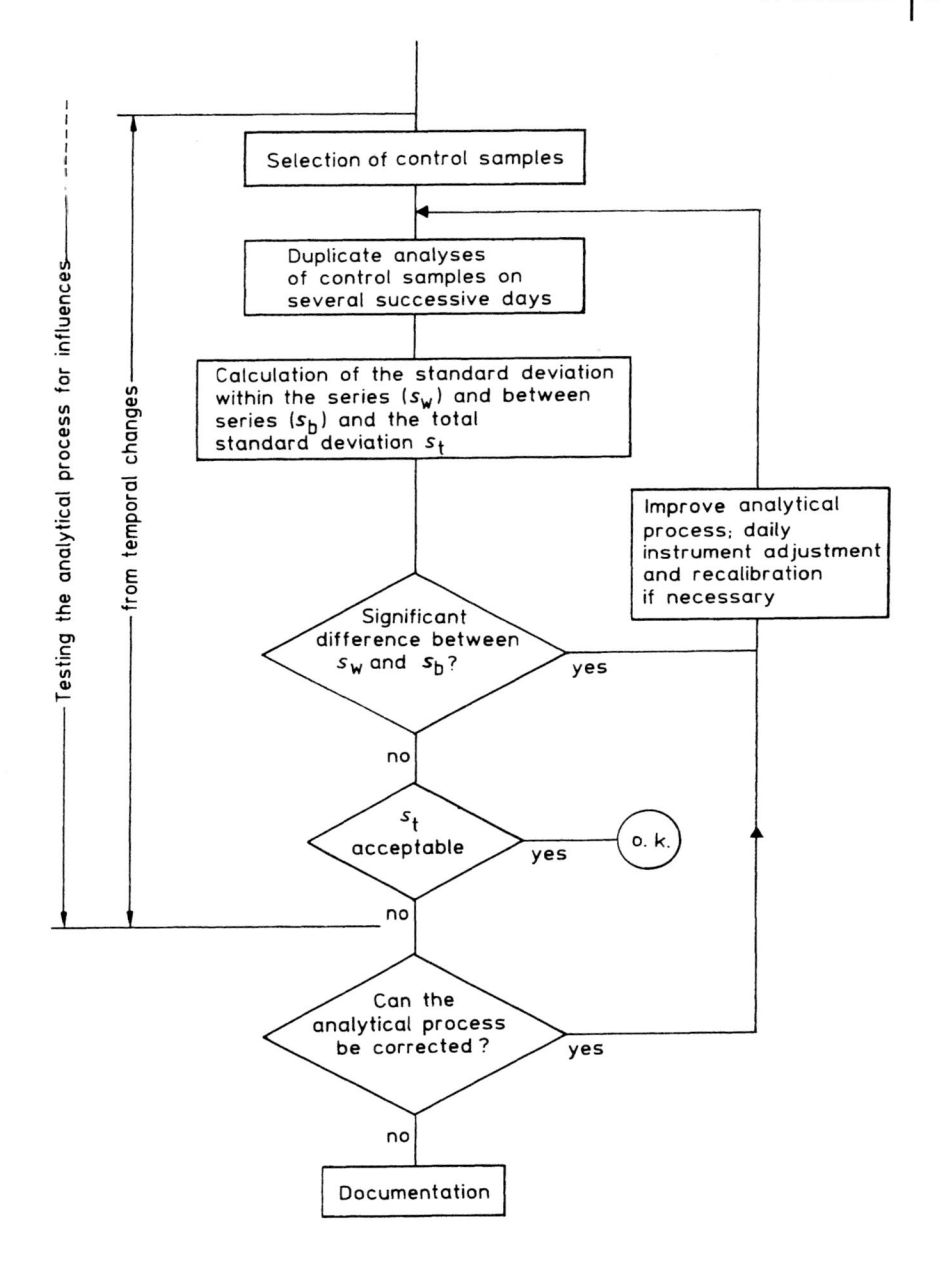

```
                    │
                    ▼
          ┌──────────────────────┐
          │ Selection of control samples │
          └──────────────────────┘
                    │
                    ▼
          ┌──────────────────────┐
          │ Duplicate analyses       │
          │ of control samples on    │
          │ several successive days  │
          └──────────────────────┘
                    │
                    ▼
          ┌─────────────────────────────────┐
          │ Calculation of the standard deviation │
          │ within the series ($s_w$) and between │
          │ series ($s_b$) and the total          │
          │ standard deviation $s_t$              │
          └─────────────────────────────────┘
```

Labels on left (vertical): Testing the analytical process for influences — from temporal changes

- Significant difference between s_w and s_b? — yes → Improve analytical process; daily instrument adjustment and recalibration if necessary
- no ↓
- s_t acceptable — yes → o. k.
- no ↓
- Can the analytical process be corrected? — yes → (loop back)
- no ↓
- Documentation

Phase I consists primarily of the following five main steps (see Scheme 1-1):

- preparation of the calibration experiment,
- testing for
  - linearity of the calibration function
  - precision (variance homogeneity, absence of outliers, securing the lower bounds of the range)
- final determination of the calibration data,
- testing for the influence of the individual process steps or for matrix effects on the calibration data,
- testing for the influence of time on the analytical process.

Of special importance is the carefully conducted correction of the analytical process during the testing phase the moment the test results of a part of the system indicate an unacceptable analytical quality. If all possibilities for improving quality within the framework of the analytical process are exhausted and have not led to an improvement, in most cases a restriction of the range (reduction of the highest substance concentration chosen) can lead to an acceptable precision and to linearity of the calibration function as well as to the required variance homogeneity.

1.1.4
Results of Phase I; Statistical Data

Phase I of analytical quality assurance lays the foundation for later routine analysis by providing quality data. Above all, this includes:

- the range tested,
- the coefficients of the calibration function:
  - in the case of a first-order calibration function ($y = a + bx$): axis intercept a and slope b (characteristic of the sensitivity of the analytical procedure),
  - in the case of a second-order calibration function ($y = a + bx + cx^2$): axis intercept a, coefficient b of the linear term, as well as the coefficient c of the quadratic term; the sensitivity E of the analytical process determined from the function,
- the standard deviation of the procedure, s_{xo}, as an absolute measure of precision for the calibration, and
- the process *variation coefficient*, V_{xo}, as a *relative* measure of precision.

In addition, the general evaluation of the analytical process also documents the following:

- *decision limit*, x_{DL}, as the substance content that produces a measurement larger than the blank value with a probability of error α (e.g., $\alpha = 5\%$),
- *minimum detectable value*, x_{MDV}, as the substance content that is larger than the blank value with a probability of error β (e.g., $\beta = 5\%$),
- *limit of quantification*, x_{LQ}, as the substance content that can be determined with a maximum allowable relative result uncertainty,

- *auxiliary test value*, x_a, for the validation of a suitable range,
- recognized constant and/or proportional systematic deviations with certain matrices, as well as an indication of the time dependency of the analytical results, the time required, and other special remarks.

1.2 Calibration of the Fundamental Analytical Procedure (Fundamental Calibration)

For analytical processes in need of or capable of being calibrated, the application of physical measurement principles does not lead directly to an analytical result; the observations only represent the result of a physical measurement, which must be converted into the analytical result using data obtained empirically using a calibration experiment [101].

The use of an *analysis function*

Analytical result = function of the observation

or in mathematical notation:

$$\hat{x} = f(\hat{y})$$

(with $\hat{y}$ as the observation and $\hat{x}$ as the substance content/analytical result)

is based on the application of the *calibration function* obtained from the calibration experiment

observation = function of substance content

or

$$y = f(x)$$

(with x = substance content of the standard sample and y = accompanying observation)

and the precision data therefrom for the determination of an unknown substance content in a sample. After solving for x, the *calibration function* then becomes the *analytical function*, which, after inserting the observation of the tested sample, $\hat{y}$, gives the analytical result, $\hat{x}$ [101].

A *fundamental calibration* represents the calibration of the fundamental analytical procedure, or, in other words, no sample preparation steps such as extraction or work-up are performed, only standards in pure solvents (e.g., distilled water) are analyzed.

1.2.1
Establishment of an Analytical Range

Every calibration begins with the choice of a preliminary range as determined by:

a) The objective of the application as related to practical conditions. This should cover a large range of applications. In addition, the mean of the range should be roughly equal to the sample concentration most often expected, as far as this is possible in individual cases.

b) The technical possibilities.
   b1) The measured values at the lower end of the range must be significantly different from the process blank values. A lower range limit is only useful when it is at least the same or larger than the *minimum detectable value* (see Section 1.3.2) of the procedure. In addition, dilution and concentration steps must be easily and flawlessly realized.
   b2) The required analytical precision must be achievable throughout the entire range (see also "limit of quantification", Section 1.3.3). Since the imprecision of an analysis increases absolutely with increasing substance content, the range in question must not be chosen too large. If another range is necessary for routine analysis, then this should be divided into overlapping segments.

The applicability of the simple linear regression equation also requires that:

   b3) The analytical precision must be constant over the entire range [151] (homogeneity of variances). Inhomogeneity of variances that is ignored can result in a large increase in the measurement uncertainty of analytical results obtained using the calibration function (see Section 1.2.4.2.1).
   b4) There must be a linear relationship between substance content and measured value (linearity of the calibration function; see Section 1.2.4.1).

In the case of inhomogeneity of variances or nonlinearity, the chosen range must be reduced so as to fulfill these conditions, or more complicated calibration methods must be chosen, for example higher order regression functions (see Section 1.2.3.2) [72, 79] or weighted regression equations [16, 159].

1.2.2
Preparation of Standard Samples

Requirements of a standard sample are:

- purity, either lack of a matrix or a defined matrix,
- homogeneity,
- representativeness: the substance to be analyzed must be present in the standard sample in ways comparable to those expected in later analysis samples; in other words, compounds must be
  – chemically similar,
  – have the same oxidation state, etc.

- stability, ability to be preserved,
- storage: samples must not be influenced by containers or outside conditions.

Preparation of a standard sample:

- During preparation of a standard sample, the precision of the balance and volumetric equipment must be taken into account. Of the two, weighing is the more exact form of sample preparation and is therefore preferable to volumetric measurement [107]. To ensure precision, these instruments must be regularly tested and calibrated.
- There should be no successive dilutions since this entails the risk of error propagation.

After establishing the preliminary range, $N = 5...10$ standard samples are prepared so that their concentrations are distributed as equidistantly as possible over the entire chosen range.

1.2.3
Determination of the Calibration Function and Process Data

The preliminary first- and second-order calibration functions are calculated from the measured values obtained from these standard samples. The process data are necessary for further statistical tests.

Notes:

1. For reasons of clarity, the physical units of the measurement signals, concentrations, and statistical data below have been omitted during the calculations and have been added only at the final result. The plausibility of the physical units of a resulting value can, if necessary, be checked by means of a *dimensional analysis*.

    Example:
    slope of the linear calibration function:
    substance content x: mg/l
    observation: peak height y: mm
    slope of the calibration function:
    [peak height per substance content] = $\dfrac{mm}{mg/l}$

    Verification by calculation formulae:

    $$\text{slope} = \frac{\sum (x_i - \bar{x}) \cdot (y_i - \bar{y})}{\sum (x_i - \bar{x})^2}$$

    $$= \frac{\sum \left\{ \frac{mg}{l} \cdot mm \right\}}{\left\{ \left(\frac{mg}{l}\right)^2 \right\}}$$

18 | *1 Phase I: Establishing a New Analytical Procedure*

$$= \frac{\left\{\frac{mg}{l} \cdot mm\right\}}{\left\{\frac{mg}{l} \cdot \frac{mg}{l}\right\}}$$

$$= \left\{\frac{mm}{mg/l}\right\} \quad \text{(q.e.d.)}$$

2. Generally, the running index for summation signs ($\sum$) has been omitted. In these cases, the "i" forms the index and it runs from 1 to N.

   Example:

   $$\sum (x_i - \bar{x})^2 \quad \text{is equivalent to} \quad \sum_{i=1}^{N} (x_i - \bar{x})^2$$

1.2.3.1 Process Data for the Linear Calibration Function

Regression analysis provides the calibration function (see Figure 1-1) with the characteristic data.

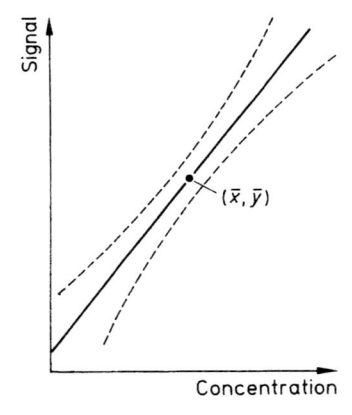

Fig. 1-1 Linear calibration function ($y = a + bx$) with prognosis interval.

Slope (measure of sensitivity):

$$b = \frac{\sum [(x_i - \bar{x}) \cdot (y_i - \bar{y})]}{\sum (x_i - \bar{x})^2} \tag{1}$$

Axis intercept:

$$a = \bar{y} - b\bar{x} \quad \text{with} \quad \bar{x} = \frac{1}{N} \cdot \sum x_i$$

$$\bar{y} = \frac{1}{N} \cdot \sum y_i \tag{2}$$

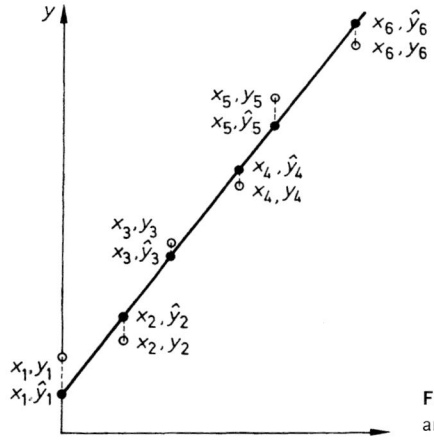

Fig. 1-2 Distribution of measured values (○) around the regression line.

Residual standard deviation (scatter of measured values around the regression line; see Figure 1-2):

$$s_y = \sqrt{\frac{\sum (y_i - \hat{y}_i)^2}{N-2}} \quad \text{with} \quad \hat{y}_i = a + bx_i \tag{3}$$

Process standard deviation:

$$s_{xo} = \frac{s_y}{b} \tag{4}$$

Process variation coefficient = relative process standard deviation:

$$V_{xo} = \frac{s_{xo}}{\bar{x}} \cdot 100\,(\%) \tag{5}$$

1.2.3.2 Process Data for the Second-Order Calibration Function

Here, regression analysis yields the second-order calibration function (see Figure 1-3) with its characteristic process data [72].

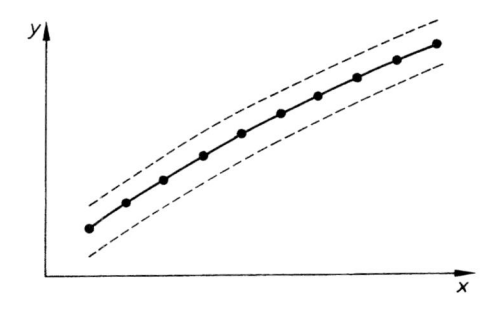

Fig. 1-3 Second-order calibration function ($y = a + bx + cx^2$).

Function coefficients a, b, c:

$$a = \frac{1}{N}\left(\sum y_i - b \cdot \sum x_i - c \cdot \sum x_i^2\right) \tag{6}$$

$$b = \frac{Q_{xy} - c \cdot Q_{x^3}}{Q_{xx}} \tag{7}$$

$$c = \frac{Q_{xy} \cdot Q_{x^3} - Q_{x^2y} \cdot Q_{xx}}{(Q_{x^3})^2 - Q_{xx} \cdot Q_{x^4}} \tag{8}$$

$$Q_{xx} = \sum x_i^2 - \left(\left(\sum x_i\right)^2/N\right) \tag{9}$$

$$Q_{xy} = \sum (x_i \cdot y_i) - \left(\left(\sum x_i\right) \cdot \left(\sum y_i\right)/N\right) \tag{10}$$

$$Q_{x^3} = \sum x_i^3 - \left(\left(\sum x_i\right) \cdot \left(\sum x_i^2\right)/N\right) \tag{11}$$

$$Q_{x^4} = \sum x_i^4 - \left(\left(\sum x_i^2\right)^2\right)/N \tag{12}$$

$$Q_{x^2y} = \sum (x_i^2 \cdot y_i) - \left(\left(\sum y_i\right) \cdot \left(\sum x_i^2\right)/N\right) \tag{13}$$

Residual standard deviation:

$$s_y = \sqrt{\frac{\sum (y_i - \hat{y}_i)^2}{N-3}} \quad \text{where} \quad \hat{y}_i = a + bx_i + cx_i^2 \tag{14}$$

Sensitivity E:

The measure of sensitivity results from the change in the measured value caused by a change in the concentration values. If the calibration function for an analytical procedure is linear, then the sensitivity is constant over the entire range and is equivalent to the regression coefficient b [72]. In the case of a curved calibration function, the sensitivity is still dependent on the given concentration value and is equivalent to the first derivative of the calibration function:

$$E(x) = b + 2c \cdot x \tag{15}$$

As a characteristic process quantity, it is recommended that the sensitivity is expressed in the middle of the working range:

$$E(\bar{x}) = b + 2c \cdot \bar{x} \tag{16}$$

From this, one can derive the process standard deviation and the relative process standard deviation:

Process standard deviation:

$$s_{xo} = \frac{s_y}{E(\bar{x})}$$

Relative process standard deviation:

$$V_{xo} = \frac{s_{xo}}{\bar{x}} \cdot 100\,(\%) \tag{17}$$

1.2.3.3 Calculating Analytical Results with the Aid of the Calibration Function

1.2.3.3.1 Results Obtained Using the Linear Calibration Function

Analytical results and confidence intervals can be calculated through the use of the linear calibration function as follows:

Result $= \hat{x} \pm CI(\hat{x})$

$$= \frac{\hat{\bar{y}} - a}{b} \pm s_{xo} \cdot t(P=95\%, f=N_c - 2) \cdot \sqrt{\frac{1}{N_c} + \frac{1}{N_a} + \frac{(\hat{\bar{y}} - \bar{y})^2}{b^2 \cdot Q_{xx}}} \tag{18}$$

with $\hat{\bar{y}}$ = mean value from N_a multiple analyses (N_a can also be equal to 1) and the data from the linear calibration:

a = axis intercept
b = slope
s_{xo} = standard deviation of the method
N_c = number of calibration standards
Q_{xx} = the sum of $(x_i - \bar{x})^2$

Example:

1. Calibration data of nitrite determination

i	x_i in mg/l	y_i in abs.
1	0.05	0.140
2	0.10	0.281
3	0.15	0.405
4	0.20	0.535
5	0.25	0.662
6	0.30	0.789
7	0.35	0.916
8	0.40	1.058
9	0.45	1.173
10	0.50	1.303

$\bar{x}$ = 0.275 mg/l
$\bar{y}$ = 0.726 abs.
a = 0.018 abs.
b = 2.575 abs./(mg/l)
s_{xo} = 0.0020 mg/l
N_c = 10
Q_{xx} = 0.20625 (mg/l)^2

2. Analytical result

The measurement obtained from an unknown sample is $\hat{y} = 0.641$ abs.
According to Eq. (18), the analytical result can be calculated with a 95% prognosis interval $[t(f=8, P=95\%) = 2.31]$:

$$\hat{x}_{1,2} = \frac{0.641 - 0.018}{2.575} \pm 0.0020 \cdot 2.31 \sqrt{\frac{1}{10} + \frac{1}{1} + \frac{(0.641 - 0.7262)^2}{2.575^2 \cdot 0.20625}}$$

$$\hat{x}_{1,2} = 0.24 \pm 0.005 \text{ mg/l}$$

1.2.3.3.2 Results Obtained Using the Linear Second-Order Calibration Function

Analytical results and confidence intervals can be calculated through the use of the linear second-order calibration function as follows:

Result for negative curvature: $\quad \hat{x} = -\dfrac{b}{2c} - \sqrt{\left(\dfrac{b}{2c}\right)^2 - \dfrac{a - \hat{y}}{c}}$

Result for positive curvature: $\quad \hat{x} = -\dfrac{b}{2c} + \sqrt{\left(\dfrac{b}{2c}\right)^2 - \dfrac{a - \hat{y}}{c}}$

with

$$CI(\hat{x}) = \frac{s_y \cdot t}{(b + 2c\hat{x})} \cdot \sqrt{\frac{1}{N_c} + \frac{1}{N_a} + \frac{1}{Q_{x^4} \cdot Q_{xx} - (Q_{x^3})^2}} \cdot$$

$$\cdot \left\{ (\hat{x} - \bar{x})^2 Q_{x^4} + \left(\hat{x}^2 - \frac{\sum x_i^2}{N_c} \right)^2 Q_{xx} - 2 \cdot (\hat{x} - \bar{x}) \cdot \left(\hat{x}^2 - \frac{\sum x_i^2}{N_c} \right) \cdot Q_{x^3} \right\} \quad (19)$$

Example:

1. Calibration data

i	x_i in mg/l	y_i in abs.
1	12	0.083
2	18	0.123
3	24	0.164
4	30	0.203
5	36	0.240
6	42	0.273
7	48	0.303
8	54	0.334
9	60	0.364
10	66	0.393

$\bar{x}$ = 39 mg/l
a = 0.00562 abs.
b = 0.00767 abs./(mg/l)
c = -0.000025 abs./(mg/l)^2
s_y = 0.00148 abs.
s_{xo} = 0.258617 mg/l
N_c = 10
Q_{xx} = 2970 (mg/l)^2
Q_{x^3} = 231660 (mg/l)^3
Q_{x^4} = 18753770 (mg/l)^4

2. Analytical result

The measured value of an unknown sample is $\bar{y} = 0.223$ abs. According to Eq. (19), the analytical result can be calculated as

$$\hat{x} = 33.46 \text{ mg/l}$$

Accordingly, the 95% prognosis interval with $t(f = 7, P = 95\%) = 2.36$ will be calculated as:

$$CI(\hat{x}) = \pm 0.643 \text{ mg/l}$$

1.2.4 Verification of the Fundamental Calibration

1.2.4.1 Verification of Linearity
If possible, one should attempt to work with a first-order calibration function. Second-order calibration functions should only be used in justified exceptions.

1.2.4.1.1 Visual Linearity Test
The type of calibration function is most simply determined through graphical representation of the calibration data, including a calibration line and a subjective evaluation. If the measured data were highly precise (see Figure 1-4) and the result is obviously nonlinear, one may do without a special statistical test of linearity. In doubtful cases, however, the linearity should be verified mathematically.

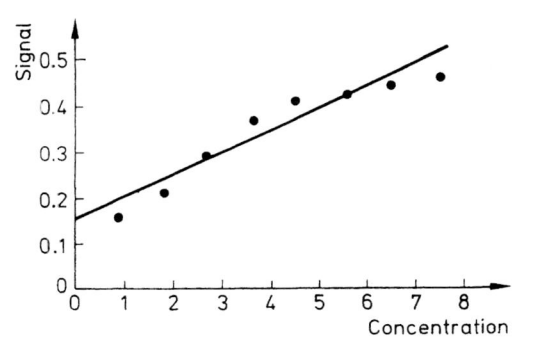

Fig. 1-4 Visual linearity test.

1.2.4.1.2 Mandel's Fitting Test
Mandel's fitting test is recommended for mathematical verification of linearity [47, 151, 190].

The test is based on the assumption that relatively large deviations of measured values from a straight line are caused by nonlinearity and may be reduced through the selection of a "better" regression model, in this case a second-order function.

For this, the first-order calibration function $y = a + bx$ and the second-order calibration function $y = a + bx + cx^2$, including their respective residual standard deviations s_y (see Sections 1.2.3.1 and 1.2.3.2), are used.

The difference of the variances DS^2 is calculated using the residual standard deviation s_{y_1} (from the first-order calibration function) and s_{y_2} (from the second-order calibration function):

$$DS^2 = (N-2)\, s_{y_1}^2 - (N-3)\, s_{y_2}^2, \text{ with the degree of freedom } f = 1 \tag{20}$$

The test value, TV, is calculated for the F-test:

$$TV = \frac{DS^2}{s_{y_2}^2} \tag{21}$$

and is compared with the value obtained from the table $F(f_1 = 1, f_2 = N-3, P = 99\%)$.

If $TV \leq F$, then the second-order calibration function will *not* provide a significantly better fit; the calibration function is linear.

If $TV > F$, then the individual steps of the analytical process should first be checked and improved upon if possible. If after that linearity is still not obtained (e.g., due to physicochemical laws), then a narrowing of the range should be attempted in order to maintain sufficient linearity. A last resort is the future evaluation of the measured data by way of a second-order calibration function (see Section 1.2.3.2). Residual analysis is an alternative to the Mandel fitting test for the verification of linearity.

1.2.4.1.3 Residual Analysis

Another possibility for testing whether the chosen functional approach of the calibration model adequately describes the measured results is residual analysis [113].

The residuals d_i are the vertical distances of the observations from the regression curve (see Figure 1-2)

$$d_i = \hat{y}_i - y_i \quad \text{for} \quad i = 1, \ldots, n$$

with y_i = observation and $\hat{y}_i$ = estimated value of y_i (from the regression function).

The residuals d_i are distributed normally [69, 113] if the chosen model approach is correct (see Figure 1-5 a). If the residuals show a trend (see Figure 1-5 b–d), then the underlying regression approach must be verified; for example, in case (d) of Figure 1-5, a second-order function must be calculated.

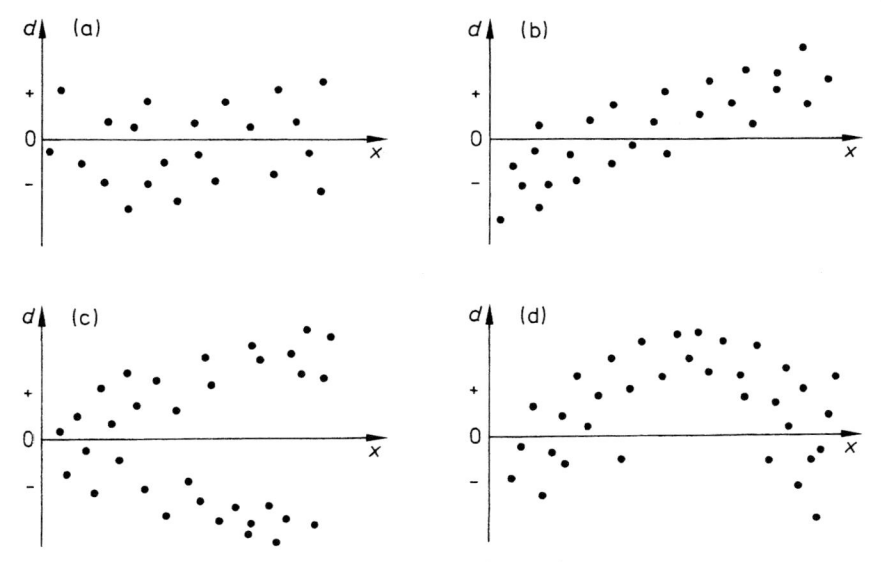

Fig. 1-5 Graphical representation of residuals dependent on the concentration x:
a) ideal course, i.e., model approach is correct;
b) linear trend, probably incorrect approach or calculation error;
c) increasing variances, i.e., inhomogeneity of variances;
d) nonlinear course, result of choosing an incorrect regression function.

1.2.4.2 Verification of Precision

1.2.4.2.1 Homogeneity of Variances

The described linear regression calculation assumes a constant (homogeneous) imprecision (variance of measured values) over the range.

Inhomogeneity leads not only to a higher imprecision (see Figure 1-6), but also to a higher inaccuracy through possible change in the linear slope.

In order to verify the homogeneity of variances, $n = 10$ standard samples of each of the lowest (x_1) and the highest (x_N) concentrations of the preliminary range are analyzed separately.

One obtains $2 \cdot n$ ($n = 10$) measurements (y_{ij}) from this series. The means, $\bar{y}_1$ and $\bar{y}_N$, and variances, s_1^2 and s_N^2, are calculated for both sets of data:

$$s_i^2 = \frac{\sum (y_{ij} - \bar{y}_i)^2}{n_i - 1} \quad \text{(for } i=1 \text{ and } i=N, \text{ and with } j \text{ from 1 to 10 in each case)} \quad (22)$$

The variances of both series of measurements are checked for homogeneity using an F-test:

$$TV = \frac{s_N^2}{s_1^2} \qquad (23)$$

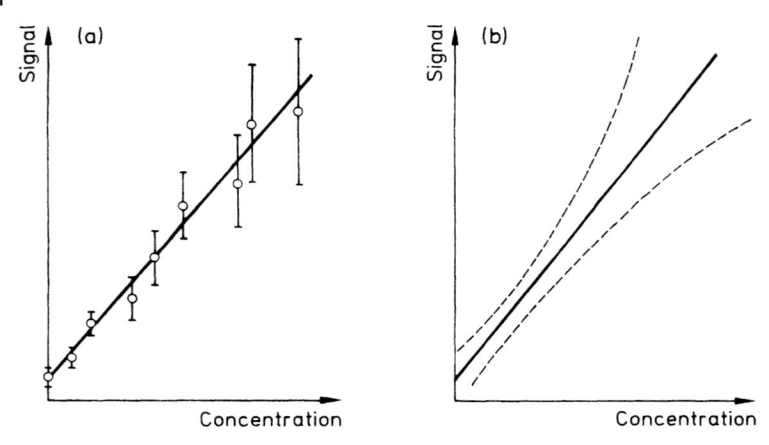

Fig. 1-6 Inhomogeneity of variances.
a) mean values and confidence intervals of the measured results
   (multiple analyses of the standards);
b) resulting prognosis bands.

If the F-test shows a significant difference between the variances, i.e., $TV > F$ ($f_1 = 9, f_2 = 9, P = 99\%$), there are three possible ways to proceed:

1. Choose a narrower range and repeat the verification for the homogeneity of variances (*recommended method*). A typical cause of the inhomogeneity of variances lies, for example, in a change in the display or amplification range of the measuring instrument within the range being used. A display change alone can affect the variance by around a factor of ten.
2. Use a weighted regression [159].
3. Multiple curve fitting [160].

1.2.4.2.2 Outlier Test

As a matter of principle, calibration data must be free from outliers. Suspected outlier values can be tested by means of various outlier tests [149]. In each case, the suitable regression model must be determined beforehand (see Section 1.2.4.1), since the correctness of the chosen regression approach is a prerequisite for the application of outlier tests.

Residual analysis (see Section 1.2.4.1.3) can also be used to determine outliers during calibration [115]. To this end, the calibration curve with the residual standard deviation must be calculated using all data pairs. Potential outliers can be pre-selected mathematically through determination of the residuals ($y_i - \hat{y}_i$; see Figure A5; Appendix A1.2.4) and their graphical representation. Every data pair with a noticeably large residual is a potential outlier. After elimination of the suspect outlier pair (x_A, y_A) from the collected data, a new calibration line is calculated with the residual distribution $s_{y_{A2}}$. Either an F-test or a t-test may be used for verification. Both methods will give identical results.

F-Test: The residual distributions $s_{y_{A1}}$ and $s_{y_{A2}}$ of the two lines are checked for a significant difference.

The test value is calculated

$$TV = \frac{(N_{A1} - 2) s_{y_{A1}}^2 - (N_{A2} - 2) s_{y_{A2}}^2}{s_{y_{A2}}^2} \tag{24}$$

and compared with the value obtained from the table F ($f_1 = 1$, $f_1 = N_{A2} - 2$, $P = 99\%$). If $TV < F$, then with a 1% error probability no outlier exists and the eliminated values can be reincluded in the collected data.

t-Test: For the *t*-test, the prediction interval of the second regression line is calculated (after eliminating the outlier) for the concentration x_A and it is checked to see if the suspected outlier value lies within this prediction interval. If this is the case, then the eliminated value must be reincluded in the collected data.

Calculation of the prediction interval:

$$CI(\hat{y}_A) = \hat{y}_A \pm t \cdot s_{y_{A2}} \cdot \sqrt{1 + \frac{1}{N_{A2}} + \frac{(x_A - \bar{x})^2}{\sum x_i^2 - \frac{1}{N_{A2}}\left(\sum x_i\right)^2}}$$

$$= a_2 + b_2 \cdot x_A \pm t \cdot s_{y_{A2}} \cdot \sqrt{1 + \frac{1}{N_{A2}} + \frac{(x_A - \bar{x})^2}{\sum x_i^2 - \frac{1}{N_{A2}}\left(\sum x_i\right)^2}} \tag{25}$$

t = value of *t*-distribution obtained from the table ($P = 95\%$, $f = N_{A2} - 2$, double-sided)
N_{A2} = $N - 1$ (N = original number of calibration data pairs)
x_A = standard concentration of the eliminated outlier observation
$\bar{x}$ = mean of all x_i (without x_A)

Note: If an outlier is statistically proven by means of an *F*- or *t*-test, then it is absolutely necessary to seek and eliminate the source of error and then repeat the entire calibration.

1.2.4.2.3 Securing the Lower Range Limit

A calibration function is only usable for quantitative analysis when all analytical results subsequently calculated with it are significantly different from zero. Therefore, the lower range limits are tested to determine if they are significantly different from zero. The calculation formulae for the test value x_a (see Figure 1-7) correspond to those used to determine the minimum detectable value (see Section 1.3.2.2).

$$x_a = 2\, CI_x\,(y = y_a) \quad \text{with } t\,(f = N - 2,\, P = 95\%,\, \text{single-sided}) \tag{26}$$

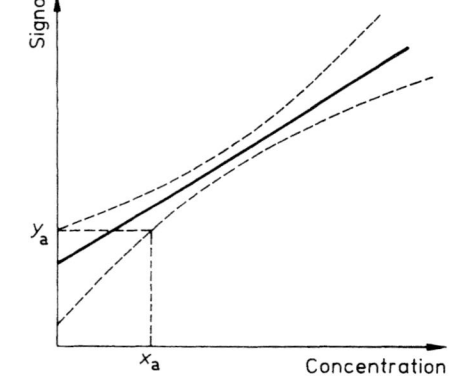

Fig. 1-7 Determination of x_a using the auxiliary value y_a.

$$x_a = 2 \cdot \frac{y_a - a}{b} \tag{27}$$

where

$$y_a = a + s_y \cdot t \cdot \sqrt{\frac{1}{N} + 1 + \frac{\bar{x}^2}{\sum (x_i - \bar{x})^2}} \tag{28}$$

If $x_a < x_1$ (see Figure 1-8a), then the entire chosen range is statistically sound, i.e. the lower limit of the range x_1 is significantly different from the concentration zero (see Section 1.3.2, minimum detectable value).

If x_a lies above x_1 (see Figure 1-8b), then x_1 is not significantly different from the concentration zero and the range is therefore only sound for concentrations $> x_a$. In this case, quantitative analyses are only possible above this concentration value. A completely new calibration must be performed for this limited range. However, it makes more sense to examine and improve the analytical process, or the individual procedures therein.

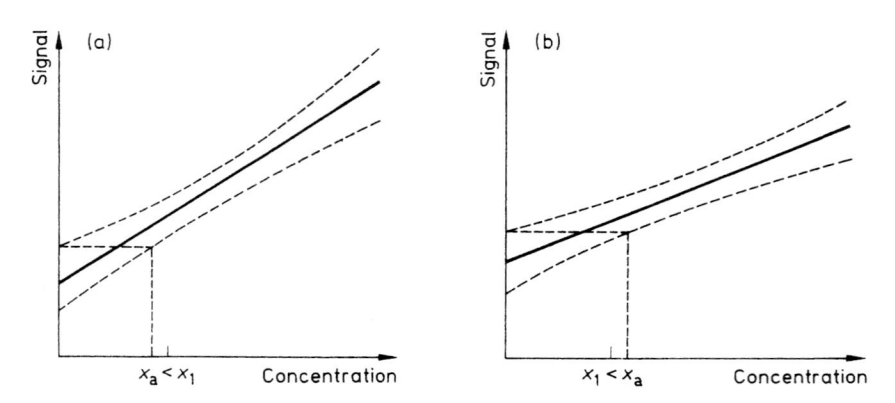

Fig. 1-8 Testing the lower range limit x_1. a) $x_a < x_1$; b) $x_a > x_1$.

1.2.4.2.4 Relative Analytical Imprecision

If a minimum precision $CI_{\text{rel,exp}}$ (relative, expected) is required for the analysis of the sought parameters, then the *relative analytical precision* at the lower end of the range (x_1) must be verified:

$$CI(x_1) = s_{xo} \cdot t \cdot \sqrt{\frac{1}{N} + 1 + \frac{(x_1 - \bar{x})^2}{\sum (x_i - \bar{x})^2}} \qquad (29)$$

$$CI_{\text{rel}}(x_1) = \frac{CI(x_1)}{x_1} \cdot 100\,(\%) \qquad (30)$$

If $CI_{\text{rel}}(x_1)$ is larger than $CI_{\text{rel,exp}}$, then either the analytical precision must be improved (optimization of individual procedure steps) or the lower range limit must be raised.

1.3
Analyses at Very Low Concentrations

In most analytical processes, the absolute precision of the measured data, and/or the analytical result, improves with decreasing concentration of the substance being analyzed. However, when working with extremely low concentrations (trace analysis) it is often a problem to significantly distinguish the signal of a sample from the signal of a blank or to obtain a quantitative result with sufficient precision.

If during an analysis an observation is obtained that lies negligibly above zero, then two contrasting interpretations are possible:

a) The sample does not contain the substance in question (analyte); the observation is only a result of the imprecision of the analytical procedure and is therefore included in the distribution range of the blank values.

b) The analyte is actually present in the sample; repeated analyses would in this case provide an analytical mean within the distribution range of which the first, and doubtful, observation would lie.

Both the probability of affiliation to the blank measurement values and also to an existing substance concentration can be statistically determined for each measured value at a known analytical precision (standard deviation). Decision-making criteria for or against the given probability of error are:

a) The probability that the sample really does not contain the analyte, even though a positive measurement has been obtained; this is known as α-error (false positive decision).

b) The probability that the tested sample really does contain the analyte, even though the analytical results are seen as zero (blank value); this is known as β-error (false negative decision), see Figure 1-9.

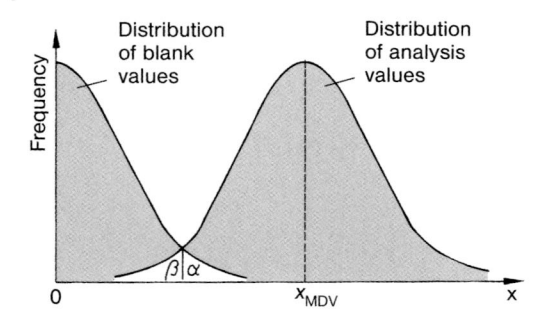

Fig. 1-9 Distribution of blank and analytical values with α- and β-errors.

In analytical practice, three characteristic values with different information content can be defined at a given precision:

- the decision limit x_{DL},
- the minimum detectable value x_{MDV},
- the limit of quantification x_{LQ}.

One must differentiate between two different courses of action for determining the decision limit and minimum detectable value depending on whether or not the analytical process is capable of being calibrated (see Figure 1-10).

In analytical processes that can be calibrated, the decision limit and the minimum detectable value will be derived from the calibration data. Therefore, a calibration can be carried out using a very low concentration that is close to the expected decision limit, minimum detectable value, and limit of quantification. This calibration range extends over at most one order of magnitude so that the variance homogeneity is given. It is usually situated well below the range for later routine analyses.

Certain analytical procedures, for example the determination of chemical oxygen demand (COD) in water analysis according to DIN 38409–41 [51] or the adsorption of organic halogens (AOX) according to DIN EN 1485 [59], cannot be calibrated because a suitable, defined (representative) standard substance does not exist.

In these cases, neither the decision limit nor the minimum detectable value can be calculated using calibration data. Instead, these are obtained by repeated analysis of blank samples.

As one can make predictions about the efficiency [88] of the future application of the analytical procedure using the characteristics "decision limit, minimum detectable value, and limit of quantification", the number of future multiple analyses per sample as N_a in the calculations must be taken into account: for single analyses $N_a = 1$, for triple analyses $N_a = 3$, and so on.

The determination and evaluation of blank values presumes:
- that the determination of blank values and measured values for the sample are independent of each other;
- that the blank value is obtained from the entire analytical process (i.e., including sample preparation and measurement);

1.3 Analyses at Very Low Concentrations | 31

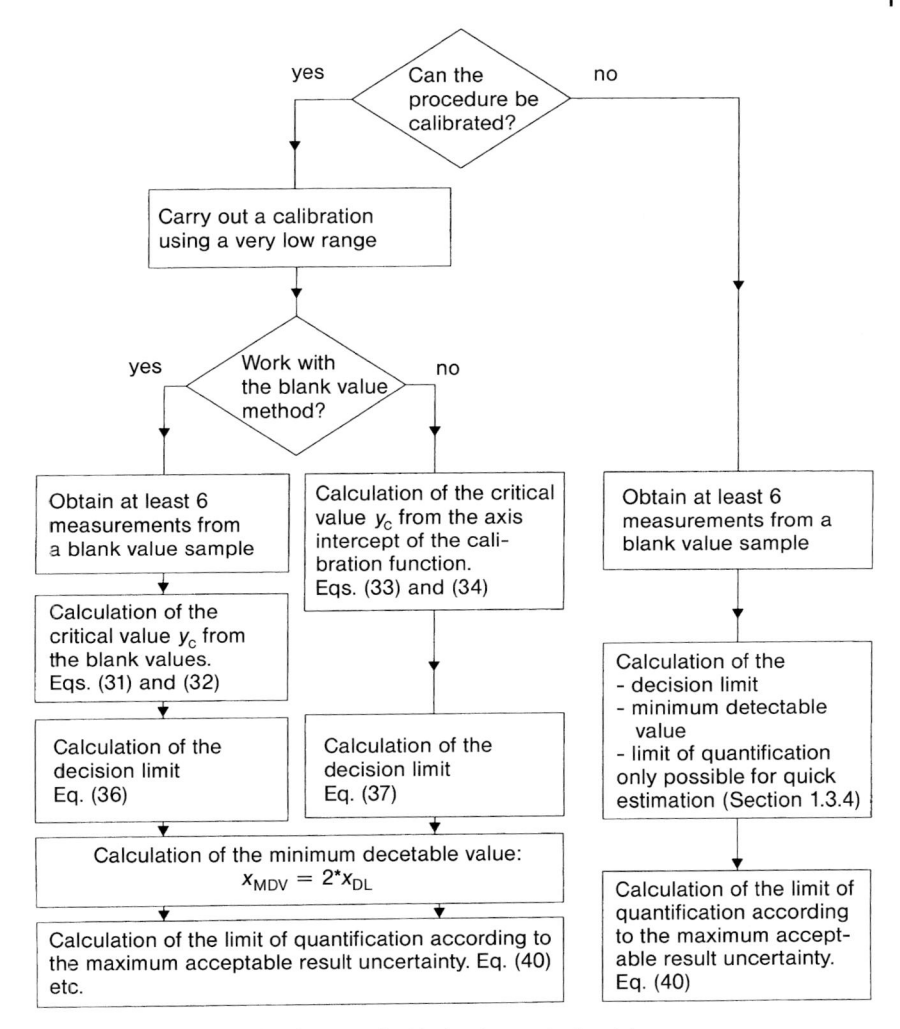

```
                    ┌─────────────┐
              yes   │  Can the    │   no
         ┌──────────│ procedure be│──────────┐
         │          │ calibrated? │          │
         │          └─────────────┘          │
         ▼                                   │
┌──────────────────────┐                     │
│ Carry out a calibra- │                     │
│ tion using a very    │                     │
│ low range            │                     │
└──────────────────────┘                     │
         │                                   │
         ▼                                   │
      ┌─────────────┐                        │
 yes  │ Work with   │  no                    │
┌─────│ the blank   │─────┐                  │
│     │ value method│     │                  │
│     └─────────────┘     │                  │
▼                         ▼                  ▼
```

yes branch (blank value)	no branch (calibration)	no (procedure not calibrated)
Obtain at least 6 measurements from a blank value sample	Calculation of the critical value y_c from the axis intercept of the calibration function. Eqs. (33) and (34)	Obtain at least 6 measurements from a blank value sample
Calculation of the critical value y_c from the blank values. Eqs. (31) and (32)		Calculation of the - decision limit - minimum detectable value - limit of quantification only possible for quick estimation (Section 1.3.4)
Calculation of the decision limit Eq. (36)	Calculation of the decision limit Eq. (37)	
Calculation of the minimum decetable value: $x_{MDV} = 2{*}x_{DL}$		Calculation of the limit of quantification according to the maximum acceptable result uncertainty. Eq. (40)
Calculation of the limit of quantification according to the maximum acceptable result uncertainty. Eq. (40) etc.		

Fig. 1-10 Criteria for deciding between the blank value method and the calibration function method.

- that the measured values of the sample and the accompanying blank values are governed by a normal distribution [190];
- that the analytical result, calculated using the difference between sample values and blank values, also follows a normal distribution; and
- that the distribution of the blank values is not significantly different from the distribution of the analytical results at low concentrations.

1.3.1
Decision Limit [34, 120, 132]

The critical value y_c for measured values is defined as that observation (measured value) for which the β-error is exactly 50% with an α-error of 5% (see Figure 1-11).

Expressed in terms of error probability, this means that if the observation value y_c is used, the blank value has been exceeded ("substance proven"), and it is subject to an error probability of only $\alpha = 5\%$. However, with repeated analysis of these samples approximately 50% of all results lie below y_c ($\beta = 50\%$), in other words, the probability that the presence of substance *cannot* be proven is 50%!

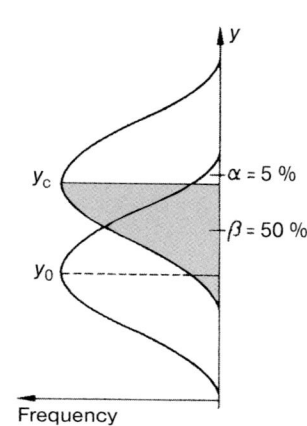

Fig. 1-11 Decision limit: α-error = 5%, β-error = 50%.

Practical Determination of the Decision Limit x_{DL}

First the critical value, y_c, is calculated. Inserting the critical value, y_c, into the calibration function and solving according to x yields the decision limit, x_{DL}.

$$x_{DL} = \frac{y_c - a}{b} \tag{31}$$

To derive y_c by blank values, the mean blank value, $\bar{y}_B$, and the prognosis interval for future blank values, $\Delta \bar{y}_B$, calculated by the standard deviation, s_B, are determined.

The mean and standard deviation should be based on at least $N_B = 6$ measured values obtained under repeat conditions.

$$\Delta \bar{y}_B = s_B \cdot t_{f,\alpha} \sqrt{\frac{1}{N_a} + \frac{1}{N_B}} \tag{32}$$

with $f = N_B - 1$ degrees of freedom gives

$$y_c = \bar{y}_B + \Delta y_B \tag{33}$$

When using the calibration line method, the critical value, y_c, is calculated as the sum of the axis intercept, a, and the predicted range of the axis intercept, Δa.

$$\Delta a = s_y \cdot t_{f,\alpha} \sqrt{\frac{1}{N_a} + \frac{1}{N_c} + \frac{\bar{x}^2}{Q_{xx}}} \tag{34}$$

with $f = N_c - 2$ degrees of freedom gives

$$y_c = a + \Delta a \tag{35}$$

After insertion of Eqs. (32) and (33) or (34) and (35) into Eq. (31), one obtains the decision limit x_{DL}:

a) Using the blank value method

$$x_{DL} = \frac{\bar{y}_B + s_B \cdot t_{f,\alpha} \sqrt{\frac{1}{N_a} + \frac{1}{N_B}} - \bar{y}_B}{b}$$

$$= \frac{s_B}{b} \cdot t_{f,\alpha} \sqrt{\frac{1}{N_a} + \frac{1}{N_B}} \tag{36}$$

with $f = N_B - 1$ degrees of freedom

b) Using the calibration line method

$$x_{DL} = \frac{a + s_y \cdot t_{f,\alpha} \sqrt{\frac{1}{N_a} + \frac{1}{N_c} + \frac{\bar{x}^2}{Q_{xx}}} - a}{b}$$

$$= \frac{s_y}{b} \cdot t_{f,\alpha} \sqrt{\frac{1}{N_a} + \frac{1}{N_c} + \frac{\bar{x}^2}{Q_{xx}}} \tag{37}$$

with $f = N_c - 2$ degrees of freedom

Since only the upper limit of the distribution range is considered here, this is called a single-sided (or one-tailed) question, and this must be kept in mind when determining the t-value from the table.

In addition, the t-factor is dependent upon both the acceptable error probability, α, and the number of degrees of freedom, f. For reasons of conformity, especially in comparing boundaries, an α of 5% should be chosen. Both the error probability, α, and the number of degrees of freedom must be stated in every case.

1.3.2
Determining the Minimum Detectable Value [34, 120]

A quantitative determination with a stated concentration is only possible when the analytical result is the same as or greater than the minimum detectable value (x_{MDV}), since only then is the required level of significance achieved (see Figure 1-12) [192].

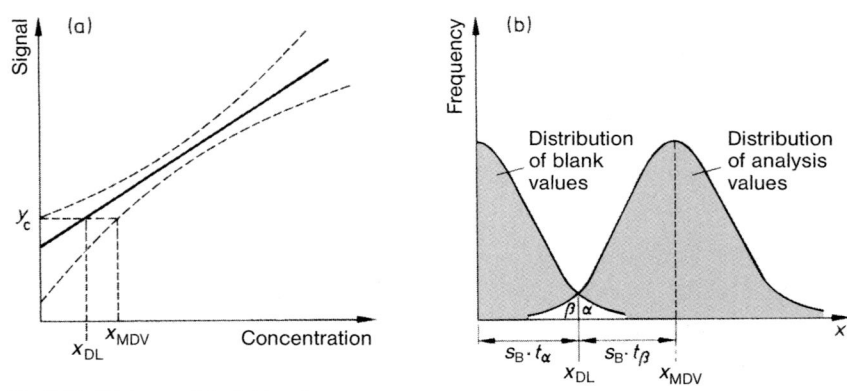

Fig. 1-12 Minimum detectable value.
a) obtained using the calibration function with its associated confidence interval;
b) obtained using the distribution of blank values; important assumption: s is identical.

This is also a single-sided (or one-tailed) question, which must again be kept in mind when determining the *t*-value from the table.

Like the decision limit, x_{DL}, the numerical value of the minimum detectable value, x_{MDV}, is dependent upon the chosen levels of significance and the number of degrees of freedom.

For equal significance levels α and β ($\alpha = \beta$), it follows that:

$$t_\alpha = t_\beta = t$$

and therefore

$$x_{MDV} = 2\, x_{DL}$$

If different significance levels are chosen, x_{MDV} must take into account the different values of *t*.

1.3.2.1 Minimum Detectable Value, Determined Using the Distribution of Blank Values

The general equation for the minimum detectable value, x_{MDV}, following the blank value method is:

$$x_{MDV} = x_{DL} + \frac{s_B}{b} \cdot t_{f,\beta} \sqrt{\frac{1}{N_a} + \frac{1}{N_B}} \qquad (38)$$

($f = N_B - 1$ degrees of freedom)

1.3.2.2 Minimum Detectable Value, Obtained Using the Calibration Function

The general equation for the minimum detectable value following the calibration method is:

$$x_{MDV} = x_{DL} + s_{xo} \cdot t_{f,\beta} \sqrt{\frac{1}{N_a} + \frac{1}{N_c} + \frac{\bar{x}^2}{Q_{xx}}} \qquad (39)$$

($f = N_c - 2$ degrees of freedom)

1.3.3 Limit of Quantification [34]

The limit of quantification is defined as the minimum concentration of a substance (x) that can be analyzed with a given maximum relative imprecision or result uncertainty, Δx_{rel}.

Example (approximately calculated):

- The determined result uncertainty is $\Delta x = 10$ mg/l.
- A relative result uncertainty, $\frac{\Delta x}{x}$, of at most 10% is assumed.
- Valid analytical results x must therefore be larger or the same as
  $$\frac{\Delta x}{10\%} = \frac{10\,\text{mg/l}}{0.1} = 100\,\text{mg/l}$$
- The limit of quantification here amounts to $x_{LQ} = x = 100$ mg/l. The limit of quantification is $k = \frac{1}{10\%} = \frac{1}{0.1} =$ ten times larger than the result uncertainty.

Table 1-1 k-Factors for the determination of the limit of quantification.

Maximum acceptable relative result uncertainty (%)	k-Factor
5	20
10	10
15	6.7
20	5
25	4
33.3	3
50	2

Calculation of the Limit of Quantification

To calculate the limit of quantification with the aid of the calibration data, α and N_a are needed in addition to the k-factor (see Table 1-1). The latter is defined as:

$$\frac{\Delta x_{LQ}}{x_{LQ}} = \text{maximum acceptable relative result uncertainty} = \frac{1}{k} \tag{40}$$

The half-width of the double-sided prognosis interval of the limit of quantification for future analyses, Δx_{LQ}, is calculated as follows:

$$\Delta x_{LQ} = s_{xo} \cdot t_{f,\alpha} \sqrt{\frac{1}{N_a} + \frac{1}{N_c} + \frac{x_{LQ} - \bar{x}^2}{Q_{xx}}} \tag{41}$$

($f = N_c - 2$ degrees of freedom)

From the definition of the relative prognosis interval (see Eq. 40):

$$x_{LQ} = k \cdot \Delta x_{LQ} \tag{42}$$

which yields the quadratic equation

$$x_{LQ} = k \cdot s_{xo} \cdot t_{f=N_c-2,\alpha} \sqrt{\frac{1}{N_a} + \frac{1}{N_c} + \frac{(x_{LQ} - \bar{x})^2}{Q_{xx}}} \tag{43}$$

A fair approximation of the limit of quantification can be obtained by multiplying the previously calculated decision limit by the k-factor: $x_{LQ} = k \cdot x_{DL}$.

The limit of quantification must *always* be expressed together with the k-value.

1.3.4
Quick Estimation

Under similar conditions ($N_c = 10$, or $N_B = 10$ and $N_a = 1$), the decision limit, minimum detectable value, and the limit of quantification can be quickly approximated using the following formulae [34]:

	for $\alpha = 5\%$	for $\alpha = 1\%$
With the blank value method:	$x_{DL} = 1.9\, s_B/b$ $x_{MDV} = 3.8\, s_B/b$ $x_{LQ} = 1.9 \cdot k \cdot s_B/b$	$= 3\, s_B/b$ $= 6\, s_B/b$ $= 3 \cdot k \cdot s_B/b$
With the calibration method:	$x_{DL} = 2.3\, s_{xo}$ $x_{MDV} = 4.6\, s_{xo}$ $x_{LQ} = 2.3 \cdot k \cdot s_{xo}$	$= 3.6\, s_{xo}$ $= 7.2\, s_{xo}$ $= 3.6 \cdot k \cdot s_{xo}$

1.3.5
Estimation of the Decision Limit and Limit of Quantification Using the S/N Ratio

Because a quick estimation using the blank value method is unsuitable for chromatographic procedures, and using the calibration procedure is often too time-consuming, in the revised version of ISO/WD 13530 [136] a simple estimation protocol for these analytical procedures is described.

The decision limit x_{DL} is defined as that concentration for which the signal-to-noise ratio (S/N) equals 3. The limit of quantification results from

$$x_{LQ} = 3 \cdot x_{DL}$$

whereby the factor 3 corresponds to a 33.3% relative result uncertainty.

1.4
Validation of Individual Process Steps and Examination of Matrix Influences

A major quality criterion of an analytical process is its applicability to complex samples. It is then expected that the developer of a method examines the process for influences such as:

- required additional steps such as sample digestion or extraction,
- interferences or matrix effects.

Typical matrices (e.g., actual surface water of typical composition) should be chosen.

Process steps and matrix effects can manifest themselves as an increase in imprecision and/or as constant systematic or proportional systematic deviations of the analytical result from the "true" value.

Calculation of the recovery function allows one to detect systematic (constant systematic as well as proportional systematic) errors, to check individual process steps, and to ascertain the influence of the matrix [100, 209].

1.4.1
Systematic Errors

1.4.1.1 Constant Systematic Errors, Additive Deviations
For constant systematic errors (see Figure 1-13), the deviation is independent of the concentration of the analyzed components, which leads to a parallel displacement of the matrix calibration line 2 (with constant systematic deviation) in relation to the calibration line 1 (prepared with pure standard solutions). The cause of this additive deviation may be the co-detection of a matrix component; the analytical procedure is therefore insufficiently specific.

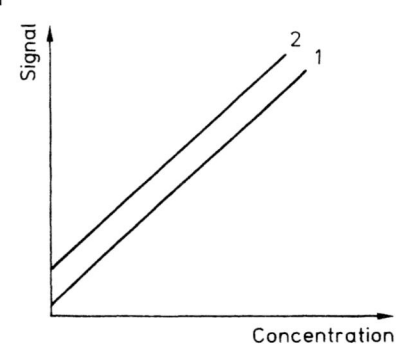

Fig. 1-13 Representation of a constant systematic deviation (no. 2).

1.4.1.2 Proportional Systematic Errors, Multiplicative Deviations

For proportional systematic deviations (see Figure 1-14), the extent of the deviation is dependent on the concentration of the analyzed components. This leads to a change in the slope of the matrix calibration line 2.

These multiplicative deviations from the true value can be the result of individual process steps (sample digestion, sample extraction, interference with frits or glassware, etc.) or matrix effects. Systematic errors can be detected by standard additions and/or the determination of the recovery function [100, 209].

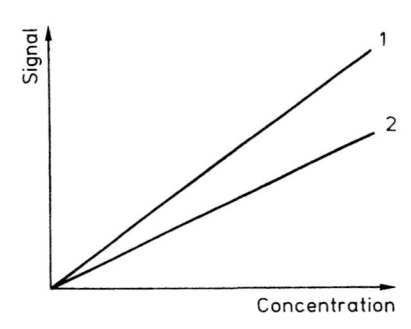

Fig. 1-14 Proportional systematic deviation (no. 2).

1.4.2
Establishment and Assessment of the Recovery Function

The objective of a recovery experiment and the establishment of a recovery function is the determination of the influence of a procedure or sample modification (also the influence of the matrix) on the analytical process. The experiments are carried out over the entire working range. Initially, the calibration function of the fundamental analytical procedure is determined:

$$y = a_c + b_c \cdot x_c \tag{44}$$

Each individual calibration sample is then subjected to the modified analytical procedure.

The analytical results x_f are then calculated using the found signal values y_f and the analysis function (the calibration function solved for x):

$$x_f = \frac{y_f - a_c}{b_c} \tag{45}$$

If the "found concentrations" (x_f) are plotted on the ordinate versus the original calibration concentrations (x_c) on the abscissa, then the *recovery curve* is obtained, which can be described mathematically by the recovery function (linear regression line):

$$x_f = a_f + b_f \cdot x_c \tag{46}$$

In the ideal case, the recovery function results in a line with intercept $a_f = 0$ and slope $b_f = 1$, as well as a residual standard deviation, s_{y_f}, that corresponds to the standard process deviation of the fundamental analytical procedure, s_{xo_c}.

1.4.2.1 Prerequisites for the Interpretation of the Recovery Function

An important prerequisite for the significance of the recovery function is the equivalence of the process standard deviation, s_{xo_c}, of the calibration function of the fundamental analytical procedure, and s_{y_f} of the calibration function of the spiked matrix or the calibration function of individual preliminary sample treatments (e.g., digestion or extraction). A matrix or a preliminary sample treatment could lead to markedly higher imprecision of the calibration (see Figure 1-15), which could possibly mask any systematic errors present.

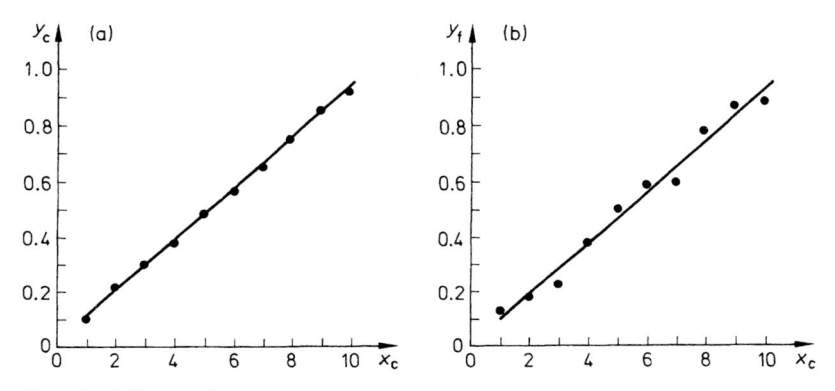

Fig. 1-15 Calibration functions
a) for aqueous standard solutions, b) for a matrix.

Checking Analytical Precision

The process standard deviation of the calibration function of the fundamental procedure, s_{xo_c}, and the residual standard deviation of the recovery function, s_{y_f}, are tested for significant difference:

$$TV = \left(\frac{s_{y_f}}{s_{xo_c}}\right)^2 \tag{47}$$

If $TV > F$ ($f_1 = f_2 = N_c - 2$, $P = 99\%$), then there is a significant difference between the two standard deviations.

In this case, no final decision with respect to the presence or absence of systematic deviations can be made. Instead, the cause of the high imprecision must be found and a new recovery function determined.

Note: The comparison of s_{y_f} and s_{xo_c} by means of an *F*-test is allowable here, since both standard deviations are calculated in concentration units.

1.4.2.2 Testing for Systematic Errors

Since measurements always contain random errors, i.e. they fall within a range, the ideal values of $a_f = 0$ and $b_f = 1$ for intercept and slope, respectively, are never obtained. In order to make a statement about the presence of systematic deviations, the confidence intervals for a_f and b_f must be determined [14].

$$CI(a_f) = a_f \pm t_{f,P} \cdot s_{a_f} \tag{48}$$

$$CI(a_f) = a_f \pm t_{f,P} \cdot s_{y_f} \cdot \sqrt{\frac{1}{N_f} + \frac{\bar{x}_c^2}{Q_{xx}}} \tag{49}$$

with

$$s_{y_f} = \sqrt{\frac{\sum [x_{if} - (a_f + b_f \cdot x_{ic})]^2}{N_f - 2}} \tag{50}$$

and N_f = the number of concentration levels.

The confidence interval for b_f can be represented by:

$$CI(b_f) = b_f \pm t_{f,P} \cdot s_{b_f} \tag{51}$$

$$CI(b_f) = b_f \pm \frac{t_{f,P} \cdot s_{y_f}}{\sqrt{Q_{xx}}} \tag{52}$$

(t = Student's *t*-factor: $f = N_f - 2$, $P = 95\%$)

The presence of systematic errors can be tested for by means of the calculated confidence interval.

If the confidence interval $CI(a_f)$ does *not* include the value $a_f = 0$, a constant systematic error is present with 95% statistical certainty.

By the same reasoning, if the confidence interval $CI(b_f)$ does *not* include the value $b_f = 1$, then a proportional systematic deviation is present with 95% certainty.

1.4.3
Application of the Recovery Function

1.4.3.1 Checking Individual Process Steps

If constant or proportional systematic errors are found during the checking of individual process steps (e.g., extraction), then the cause(s) of these errors should be sought if possible. The analytical process should then be optimized and the measurements should be repeated in order to determine the recovery function.

If systematic errors cannot be eliminated, then the process documentation must *clearly* indicate this and in practice either the calibration must be performed over the entire process (including sample preparation) or the method of standard addition must be applied.

For *exclusively* proportional systematic errors, the recovery rate (RR) can be given. This is determined from the slope b_f:

$$RR = b_f \cdot 100\,(\%) \tag{53}$$

1.4.3.1.1 Meaning of the Recovery Rate

The recovery rate is an assessment criterion for a given analytical process or individual process step [154]. If a recovery rate of 100% is obtained when verifying individual process steps, and the process in question is free from constant systematic errors, the determination of analytical results need not be performed using the method of standard addition (see Section 3.4.1.1).

However, if a proportional systematic error is found, future analytical results must be obtained using the method of standard addition; if a constant systematic error is discovered, a corresponding warning must be entered in the analysis protocol.

The recovery rate may also be used as a control quantity for a quality control chart (see Section 2.6.7.1.5.3). The central line can be chosen as either $RR = 100\%$ or as equal to the mean recovery rate. This is dependent on the results of the preliminary period.

1.4.3.1.2 Impact of a Constant Systematic Error on the Recovery Rate

Attempting to describe the trueness of an analytical process by using the recovery rate (RR) derived by replicate analyses of one sample alone leads to incorrect results if an additional constant systematic error is present (see Figure 1-16).

For every recovery function having either
- a positive intercept ($a_f > 0$) and a slope $b_f < 1$ (1)
- or a negative intercept ($a_f < 0$) and slope $b_f > 1$ (2)

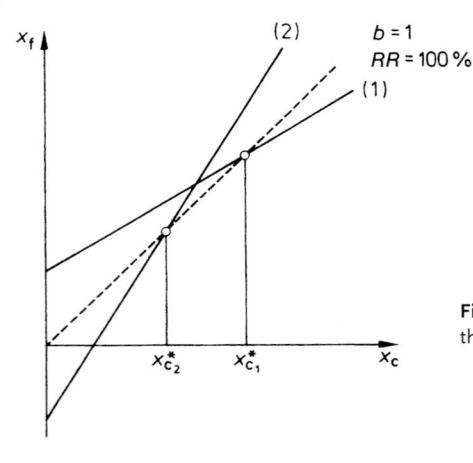

Fig. 1-16 Dependence of the recovery rate on the slope.

an RR of 100% can be found for a single concentration $x_c^* > 0$ (intersection point of the recovery curve with the angle bisect). For all concentrations $x_c \neq x_c^*$ however, the $RR \neq 100\%$.

In general, and with respect to the intercept, a_f, it follows that

$$RR = \frac{x_f}{x_c} \cdot 100\% = \left(\frac{a_f}{x_c} + b_f\right) \cdot 100\% \tag{54}$$

The RR is independent of x_c and is determined solely by b_f only if $a_f \approx 0$.

a) Proportional errors only:

Example:

x_f	$= 0.001 + 0.78 \cdot x_c$
RR	$= \left(\dfrac{0.001}{x_c} + 0.78\right) \cdot 100\%$
RR_1 ($x_c = 10$)	$= (0.0001 + 0.78) \cdot 100\%$
	$= 78.010\%$
RR_2 ($x_c = 50$)	$= (0.00002 + 0.78) \cdot 100\%$
	$= 78.002\%$
RR_3 ($x_c = 100$)	$= (0.00001 + 0.78) \cdot 100\%$
	$= 78.001\%$

b) With additional significant constant error:

Example:

x_f	$= 1.0 + 0.78 \cdot x_c$
RR	$= \left(\dfrac{1.0}{x_c} + 0.78\right) \cdot 100\%$
RR ($x_c = 10$)	$= (0.1 + 0.78) \cdot 100\%$
	$= 88\%$

$RR (x_c = 50)$ $\quad = (0.02 + 0.78) \cdot 100\%$
$\quad = \mathbf{80\%}$
$RR (x_c = 100)$ $\quad = (0.01 + 0.78) \cdot 100\%$
$\quad = \mathbf{79\%}$

Consequence: the constant systematic error a_f adds to the RR as $\dfrac{a_f}{x_c}$.

This portion of the error has its greatest value at the lower end of the range, $x_c = x_1$, and decreases hyperbolically with increasing x_c.

If, for example, the RR is allowed to be subject to a maximum inaccuracy due to a_f of 1%,

$$\frac{a_f}{x_c} \leq 0.01$$

then the constant systematic error must be $a_f \leq 0.01 \cdot x_1$.

Example of Checking a Process Step: Optimization of an Extraction Step [140]

A solid-phase extraction (using EXTRELUT® columns) was performed before the quantitative HPTLC determination of phenobarbital in urine. In order to check solely the extraction step, six standard solutions of phenobarbital were submitted to the entire extraction procedure: 5 ml standard solutions were diluted with 15 ml of sodium phosphate buffer (pH 6.0), applied to the EXTRELUT® columns, and eluted after 10 minutes with 40 ml of a mixture of a dichloromethane/2-propanol (93:3, v/v). The extracts were then concentrated to dryness in a nitrogen stream, and the residues were taken up in 500 μl of methanol and applied to HPTLC silica gel plates.

Non-extracted standards of the same concentration were also concentrated under nitrogen, and the residues were redissolved in methanol and analyzed in exactly the same way as the extracted standard solutions.

A calibration function was determined using the measured values of the non-extracted standards and their respective concentrations. The recovery function

$$x_f = -100.2 \text{ ng/spot} + 1.098 \cdot x_c$$

was obtained by linear regression of the found concentrations x_f versus the calibration concentrations x_c (see Figure 1-17).

The confidence intervals of a_f and b_f were found to be:

$$CI(a_f) = (-100.2 \pm 13.698) \text{ ng/spot}$$

$$CI(b_f) = 1.098 \pm 0.0969$$

that is, both a constant systematic and a small proportional systematic error were present.

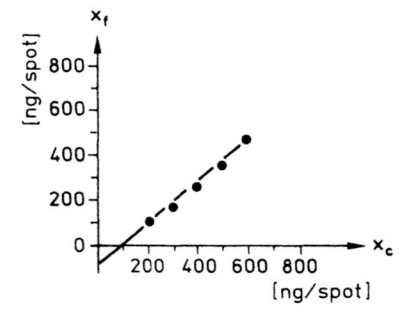

Fig. 1-17 Recovery function 1 for the phenobarbital extraction.

The cause of these systematic deviations was found to be insufficient loading of the EXTRELUT® column. Approximately 1 ml of the EXTRELUT® material at the lower end of the column remained unwetted when the column was charged with a sample volume of only 20 ml. This unwetted part of the absorbing material reabsorbed a portion of the eluted phenobarbital during elution of the column.

For this reason, an additional 1.5 ml of phosphate buffer was added to the original sample volume (20 ml) and the recovery experiment was repeated. The recovery function (see Figure 1-18)

$$x_f = 2.37 \text{ ng/spot} + 0.98 \cdot x_c$$

with the confidence intervals

$$CI(a_f) = (2.37 \pm 13.975) \text{ ng/spot}$$

$$CI(b_f) = 0.98 \pm 0.082$$

was determined. Constant or proportional systematic errors were no longer present.

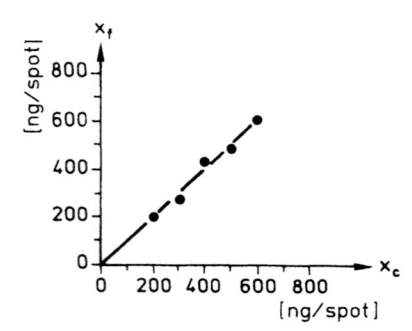

Fig. 1-18 Recovery function 2 for the phenobarbital extraction.

1.4.3.2 Determination of the Recovery Function to Prove the Influence of a Matrix

In order to prove the possible influence of a matrix, various typical sample matrices should be prepared that do not contain the analyte (possibly a synthetic sample matrix) and these should be divided into ten equal aliquots. A *concentrated* standard solution is then added to each of these ("spiking"), so that each of the matrix partial samples contains the same analyte concentration as in the aqueous solutions used in the calibration.

The "samples" thus produced are then analyzed using the appropriate analytical process. The measured values, y_f, are then converted into concentrations, x_f, using the analysis function (Eq. 45).

The recovery function as well as the residual standard deviation and the confidence interval of the intercept a_f and slope b_f are calculated and evaluated as described in Section 1.4.2.2 (see Eqs. 48 to 51).

Note: If an actual matrix that contains the analyte is used for these experiments, then it is not possible to detect the presence of a constant systematic error.

If this check for matrix influence confirms the presence of a proportional or constant systematic error, then the analysis may *not* be interpreted later using a calibration function determined with aqueous standard solutions, but the method of addition of a standard [138] must be applied in each case.

Example: HPTLC Determination of Selenium in Human Serum [100]

To determine the selenium content in human serum, a wet-chemical sample preparation is necessary. In order to determine the influence of this biological matrix on the recovery rate, the recovery function must first be determined. A serum sample is divided into six parts, five of which are spiked with various seleno-cysteine concentrations. The five spiked samples and one unspiked sample are first concentrated to dryness at 100 °C and then treated with nitric acid and hydrogen peroxide.

The individual samples are then extracted by means of an EXTRELUT® column and the extracts are applied to an HPTLC silica gel 60 plate, separated, and scanned. The following recovery function was obtained (see Figure 1-19):

$$x_f = 10.609 \text{ pg/spot} + 1.002 \cdot x_c$$

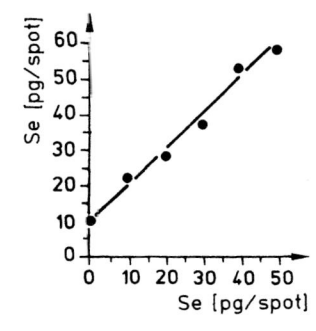

Fig. 1-19 Recovery function for selenium in human serum.

The confidence interval of the slope b_f (0.87 < b_f < 1.14) shows that no proportional systematic error is present. In other words, the determination of selenium in human serum does not have to be performed using the method of standard addition. A statement about the presence of a constant systematic error is not possible, since the matrix used for this recovery experiment was not demonstrably free of selenium.

1.5
Additional Statistical Methods

In cases of proven inhomogeneity of variances (see Section 1.2.4.2.1) as well as for disruptions in quantification, e.g. sub-optimal separation of signals (interferences, peak overlap), other more demanding and more complex statistical methods may be used [146, 153]. These include:

- weighted regression [159],
- multiple curve fitting [160],
- multiple regression [146],
- multivariate standard addition [146],
- process functions instead of process data [146].

However, these methods are beyond the scope of this book. Besides the increased calculational demands, the analytical requirements associated with these procedures are also very high.

1.6
Use of Internal Standards [50]

Multiple component procedures are analysis procedures whereby a single sample is analyzed for more than one component at a time. Gas chromatography and ICP-AES are examples. In principle, strategies involving calibration of the fundamental analytical procedure and verification by means of the recovery function, as described in the previous sections, are applicable to these procedures.

For these analyses, additional mutual influences must also be taken into account. Quality assurance measures here also include the control of equipment parameters. This is achieved, for example, through the use of internal standards.

1.6.1
Definition, Purpose

An internal standard is a substance which is known not to be present in the analysis sample and which is added in defined form and quantity to each calibration and analysis sample to be quantitatively co-analyzed. It is assumed that both the standard and the analyte are subject to the same physicochemical influences, such as the parameter to be examined. Analytical results obtained with internal standards are used:

- For checking purposes: Assuming that the internal standard is added *after all preparation steps*, immediately before the measurement, then the results obtained with an internal standard can be related to the control of the sample application (e.g., injection) or sample manipulation within the equipment, or to the detection step. In contrast, the addition of standards *before sample preparation* makes it possible to check specific preparation steps.
- Under certain preconditions, for mathematical correction of results when systematic equipment errors occur.

1.6.2
Conditions and Limitations of the Use of Internal Standards

A substance which is to be used as an internal standard must fulfill certain conditions:

- It must not, with a high degree of certainty, be present in the natural sample.
- It may not itself cause matrix effects.
- Its physicochemical properties must be as similar as possible to those of the analyte. For example, this includes a similar boiling point (in chromatography), a similar but not identical retention time, similar detectability, etc. If sample preparation steps are to be checked using internal standards, then similar chemical properties are necessary as well.
- It must be possible to add the internal standard in highly concentrated form in order to avoid volumetric errors. Otherwise, the additional volume in the sample must be considered when calculating the results of the analysis.
- The concentration of the internal standard in the sample must be suited to the measurement problem.
- A calibration and determination of the process data is obviously also necessary for an internal standard.

The use of an internal standard does not dispense with the need to perform a complete calibration on each parameter to be analyzed as well as perform experiments to determine process and matrix influences. Internal standards are also not a substitute for calibration functions.

1.6.3
Procedure

In order to monitor or correct an analytical process with the aid of one or more internal standards, it is necessary to submit the entire analytical process to a complete calibration and checking of process data. The influence of the individual process steps and matrices on the analytical process requires special scrutiny. If despite all efforts instabilities or matrix effects cannot be brought under control, then one may consider the use of one or more internal standards. Two or more internal standards may be used simultaneously when different influence quantities are to be monitored selectively. For example, internal standards of varying retention times can be used to monitor the entire separation and the stability of the detector for time-intensive chromatographic separations.

After the selection of suitable standard substances, a basic calibration is performed for these and tests are carried out to determine the influences of individual process steps and especially matrices. After the process data have been obtained for the substances to be used as internal standards, a calibration is performed on the standard mixtures. Repeated analyses of known samples complete the preliminary experiments.

If the internal standards serve only for checking purposes, then the results of these analyses form the basis of appropriate quality control systems/charts in routine analysis (see Section 2.6).

If internal standards are used for the systematic correction of results, then their recovery rates are included in each individual result [50]. The recovery rate of an internal standard is regarded as representative of the recovery rate of the analyzed substances. To obtain the corrected analytical results, the uncorrected results should be divided by the recovery rate of the internal standard.

Example:

- The uncorrected result is 50 µg/l.
- A similar substance used as the internal standard is present at a concentration of 78 µg/l. The analytical result for this internal standard is obtained as 62 µg/l, indicating a recovery rate, RR, of $\dfrac{62}{78} = 0.795$.
- Using this RR to correct the result of 50 µg/l, the adjusted concentration is

$$\frac{50\ \mu g/l}{RR} = \frac{50\ \mu g/l}{0.795} = 62.9\ \mu g/l.$$

Provided that the concentration of the internal standard is known and is constant throughout, the quotient

$$\frac{\text{measured value of parameter in question}}{\text{measured value of internal standard}}$$

can be used for the calibration and evaluation instead of the original measured values of the parameter to be examined.

1.7 Preparing for Routine Analysis

1.7.1 Examination of the Time Dependency of Measured Values

A quality objective for routine analysis is to maintain constant precision and accuracy of the results over a longer time period, or in other words to achieve a reliable analytical process.

The calibration is, as a rule, performed under so-called repetitive conditions, but nevertheless within a small time frame. The observation of analysis quality over a longer time period makes apparent a higher imprecision in the analytical result, due perhaps solely to such factors as longer pauses between individual series of analyses, environmental influences, and the varying daily "state of mind" of personnel. It may be possible that this additional imprecision prevents any kind of comparison of the analytical results. Additionally, temporal instability of analysis parameters is noticeable for some analytical procedures. The effects of aging, contamination, etc., can lead to systematically lower values within a certain operational time, even within a working day.

The objective in preparing for a routine analysis is the recognition of such effects and influences, their quantification, and, where possible, their elimination. So-called control samples (see Section 2.5) are necessary for their detection; these are samples which remain stable over longer periods of time and are analyzed as controls in each analysis series.

1.7.1.1 Comparison of the "Within Batch" Standard Deviation (s_w) with the "Between Batches" Standard Deviation (s_b) [215]

For precision analyses, the standard deviation within batch, s_w, and the standard deviation between batches, s_b, are determined and tested for significant difference using an F-test.

The determination of s_b and s_w is performed on $N = 20$ consecutive days (analysis series of a day = batch; at least 6 to 10 different analysis days or analysis series) and multiple, or at least duplicate, determinations are made for each control sample. The individual standard deviation (s_i) within batch i, the within batch standard deviation (s_w) of all batches, the between batch standard deviation (s_b), and the total standard deviation (s_t) are all calculated from the results.

The standard deviation of N_i determinations of a control sample within batch i:

$$s_i = \sqrt{\frac{1}{f_i} \cdot \sum_i (x_{ij} - \bar{x}_i)^2} \quad \text{with } f_i = N_i - 1 \tag{55}$$

for duplicate determinations: $\quad f_i = 2 - 1 = 1$

Within batch standard deviation:

$$s_w = \sqrt{\frac{1}{f_w} \cdot \sum (f_i \cdot s_i^2)} \qquad (56)$$

where $f_w = \sum f_i$; for duplicate determinations $f_w = N$ (57)

Between batch standard deviation (from batch to batch or day to day):

$$s_b = \sqrt{\frac{1}{f_b} \sum (\bar{x}_i - \bar{\bar{x}})^2} \qquad (58)$$

where $f_b = N - 1$ and $\bar{\bar{x}}$ = grand average = $\frac{1}{N} \cdot \sum \bar{x}_i$ (59)

The total standard deviation is thus given by:

$$s_t = \sqrt{\frac{1}{f_t}\left(s_b^2 \cdot f_b + s_w^2 \cdot f_w\right)} \qquad (60)$$

where $f_t = f_b + f_w$; for duplicate determinations: $f_t = 2 \cdot N - 1$ (61)

Evaluation:

1. The total standard deviation must fulfill the quality requirements (maximum allowable imprecision).

2. The between batch standard deviation should be no more than twice the within batch standard deviation; if necessary, the variance F-test may be applied to s_b and s_w in order to assess the precision:

$$\text{Calculation of the test value: } TV = \frac{s_b^2}{s_w^2}$$

Assessment of the test value: if $TV \leq F(f_b, f_w, P = 95\%)$, then s_b is only coincidentally larger than s_w.

If $TV > F(f_b, f_w, P = 95\%)$, then the imprecision between the series significantly influences the total imprecision. In the event of the total standard deviation being unacceptably high, significant quality improvements can be achieved by tracking and correcting analysis influences, and/or by readjusting or even recalibrating each series.

A Practical Example from Water Analysis

To determine dissolved organic carbon (DOC), a control sample with a content of 5 mg/l was chosen and a duplicate determination was performed on each of ten consecutive days. The results were:

$s_w = 0.099$ mg/l

$s_b = 0.224$ mg/l

$s_t = 0.158$ mg/l

$TV = \dfrac{s_b^2}{s_w^2} = 5.119$ (table value: $F\,(9, 10, 99\%) = 4.94$)

Since $TV > F\,(9, 10, 99\%)$, s_b and s_w are significantly different.

The cause of the increased imprecision is in this case an unstable calibration function. Daily recalibration could substantially improve the analytical results.

1.7.1.2 Determining the Need for Daily Adjustment of Analytical Equipment

If a result is determined to be time dependent, then a daily adjustment before each series of analyses using one or two standard solutions is recommended in order to determine the *current slope* of the calibration function. A prerequisite for daily adjustment is, however, that the process standard deviations, s_{xo}, for various slopes are not significantly different (see Figure 1-20b). This can be tested by means of an *F*-test.

For a constant residual standard deviation s_y (case (a) in Figure 1-20), before applying a daily adjustment one must determine by how much the *current slope* of the calibration function (determined using the daily adjustment) may be decreased while still allowing evaluation over the entire chosen analytical range. Verification by means of the maximum acceptable imprecision is recommended (see Section 1.3.3) [22].

If the slope of the calibration function decreases with constant residual standard deviation s_y (Figure 1-20a), then it may be that the required precision $CI_{rel,req}$ for x_1 can no longer be maintained.

If no imprecision limit is available from $CI_{rel,req}$, then the lower range limit must be tested using the test value x_a (see Section 1.2.4.2.3 and Fig. 1-21). Below the concentration x_a, no definitive analytical statement can be made. The practitioner may, however, perform quantitative analysis above this point provided that the values stay above the limit of quantification of the analytical process.

1.7.1.3 The Trend Test

The procedure described in Section 1.7.1.1 is not suitable to distinguish between a time-dependent *systematic trend* and a simple imprecision. Therefore, a simple trend test (according to Neumann [190]) should be applied.

A prerequisite for the execution of the trend test is the availability of a series of temporally successive analytical results of a control sample $x_1, x_2, ..., x_n$, which originated from a normally distributed base entity. At least $n = 20$ analysis results should be included, if possible.

In addition to the standard deviation, s, of the n values, the mean quadratic of the $n - 1$ differences of consecutive values (single or mean values of consecutive series; successive difference dispersion) Δ^2 is calculated:

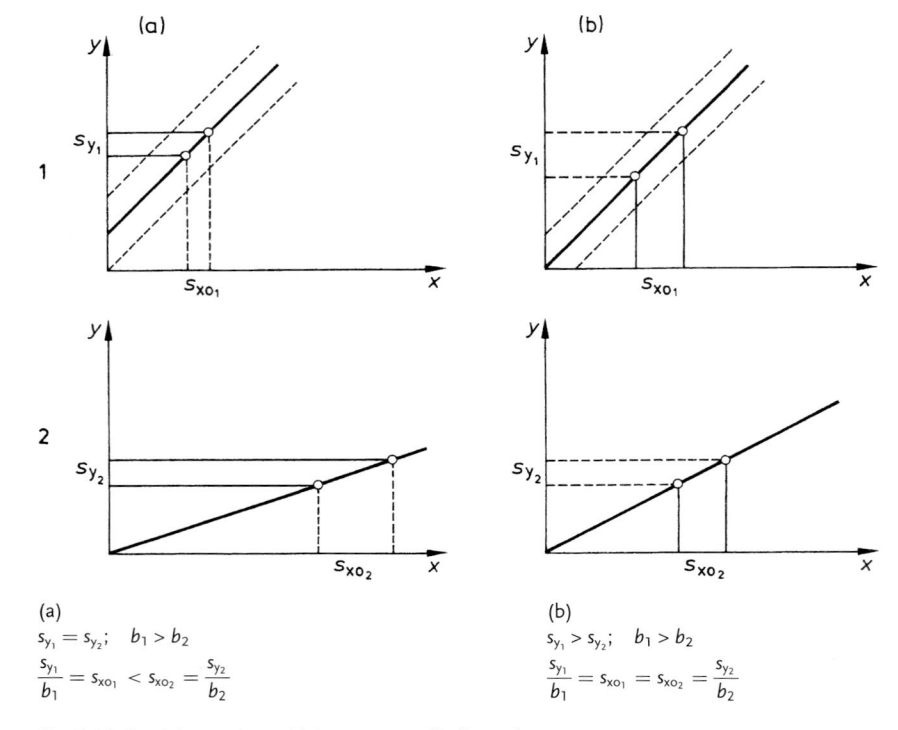

Fig. 1-20 Precision and sensitivity are mutually dependent.

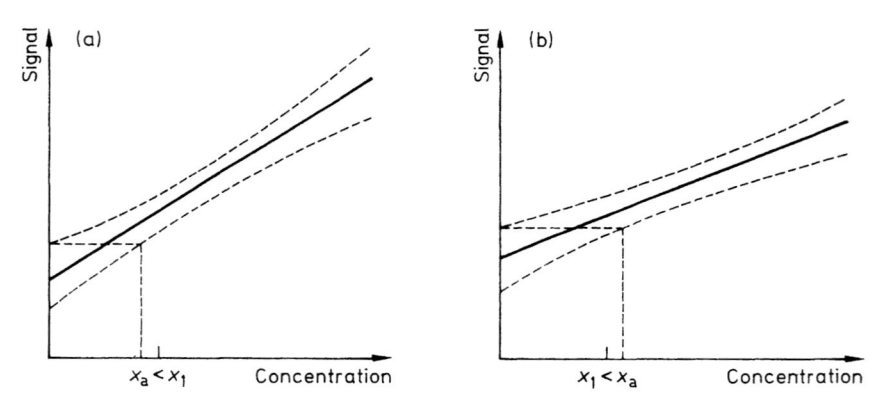

(a)
$s_{y_1} = s_{y_2}; \quad b_1 > b_2$
$\dfrac{s_{y_1}}{b_1} = s_{xo_1} < s_{xo_2} = \dfrac{s_{y_2}}{b_2}$

(b)
$s_{y_1} > s_{y_2}; \quad b_1 > b_2$
$\dfrac{s_{y_1}}{b_1} = s_{xo_1} = s_{xo_2} = \dfrac{s_{y_2}}{b_2}$

Fig. 1-21 Statistical check of a range by means of x_a for daily adjustment.
a) x_1 is significantly different from the concentration zero,
b) x_1 is *not* significantly different from the concentration zero.

1.7 Preparing for Routine Analysis

$$\Delta^2 = \frac{\sum (x_i - x_{i+1})^2}{(n-1)} \tag{62}$$

If the consecutive values are independent, then $\Delta^2 \approx 2s^2$.
The hypothesis that consecutive values are independent must be rejected in favor of the alternative hypothesis that a trend exists, if the quotient

$$\frac{\Delta^2}{s^2} = \frac{\sum (x_i - x_{i+1})^2}{\sum (x_i - \bar{x})^2} \tag{63}$$

is less than or equal to the critical limit (n, $\alpha = 1\%$) (see Table 1-2).

Table 1-2 Critical limits for the quotients from the mean quadratic successive difference dispersion and the variance [190].

n	0.1%	1%	5%	n	0.1%	1%	5%
4	0.5898	0.6256	0.7805	33	1.0055	1.2283	1.4434
5	0.4161	0.5379	0.8204	34	1.0180	1.2386	1.4511
6	0.3634	0.5615	0.8902	35	1.0300	1.2485	1.4585
7	0.3695	0.6140	0.9359	36	1.0416	1.2581	1.4656
8	0.4036	0.6628	0.9825	37	1.0529	1.2673	1.4726
9	0.4420	0.7088	1.0244	38	1.0639	1.2763	1.4793
10	0.4816	0.7518	1.0623	39	1.0746	1.2850	1.4858
11	0.5197	0.7915	1.0965	40	1.0850	1.2934	1.4921
12	0.5557	0.8280	1.1276	41	1.0950	1.3017	1.4982
13	0.5898	0.8618	1.1558	42	1.1048	1.3096	1.5041
14	0.6223	0.8931	1.1816	43	1.1142	1.3172	1.5098
15	0.6532	0.9221	1.2053	44	1.1233	1.3246	1.5154
16	0.6826	0.9491	1.2272	45	1.1320	1.3317	1.5206
17	0.7104	0.9743	1.2473	46	1.1404	1.3387	1.5257
18	0.7368	0.9979	1.2660	47	1.1484	1.3453	1.5305
19	0.7617	1.0199	1.2834	48	1.1561	1.3515	1.5351
20	0.7852	1.0406	1.2996	49	1.1635	1.3573	1.5395
21	0.8073	1.0601	1.3148	50	1.1705	1.3629	1.5437
22	0.8283	1.0785	1.3290	51	1.1774	1.3683	1.5477
23	0.8481	1.0958	1.3425	52	1.1843	1.3738	1.5518
24	0.8668	1.1122	1.3552	53	1.1910	1.3792	1.5557
25	0.8846	1.1278	1.3671	54	1.1976	1.3846	1.5596
26	0.9017	1.1426	1.3785	55	1.2041	1.3899	1.5634
27	0.9182	1.1567	1.3892	56	1.2104	1.3949	1.5670
28	0.9341	1.1702	1.3994	57	1.2166	1.3999	1.5707
29	0.9496	1.1830	1.4091	58	1.2227	1.4048	1.5743
30	0.9645	1.1951	1.4183	59	1.2288	1.4096	1.5779
31	0.9789	1.2067	1.4270	60	1.2349	1.4144	1.5814
32	0.9925	1.2177	1.4354	∞	2.0000	2.0000	2.0000

Meaning of the Trend Test

The trend test may be used to check if, compared to the within batch standard deviation (s_w), a larger standard deviation between batches (s_b, see Section 1.7.1.1) is a result of a temporal systematic deterioration of the detection system. As an example, the aging of a detector in gas chromatography can be determined in this way.

Consequence

If a temporal trend can be proven, then the quality of the analytical results can be significantly improved by readjustment or even recalibration in each series.

1.8
Summary of the Results of Phase I (Process Development): Documentation

The final documentation of a fundamental analytical process includes, in addition to the description of the analysis procedure, all relevant information about the analysis quality.

Since the quality of analytical results is closely related to the quality of the instructions, the latter should be clear, understandable, and practically oriented [17, 123] so that an unacquainted analyst can follow them to obtain satisfactory results.

The instructions are to be completed with the following information, for the most part obtained in Phase I:

- the substance to be analyzed (analyte),
- area of application, type of possible bulk (or gross) samples,
- tolerances (e.g., "add 6.0 ml ± 0.1 ml"), in order to obtain the necessary precision at each analytical step,
- measuring instructions, including physical quantity and unit of the measured value,
- the procedure to determine the calibration function and the blank value,
- the working range,
- calibration instructions: number, type, preparation of standard solutions,
- coefficients of the calibration function with information about the linearity, sensitivity, and residual standard deviation,
- the process standard deviation,
- the decision limit, with information about how it was obtained including how many measurements were used,
- the relative imprecision at the lower limit of the range ($CI_{rel}(x_1)$) for later comparison with the required limit of quantification,
- instructions for the evaluation and description of the results,
- frequently seen process errors and accompanying countermeasures,
- quantitative information about interfering substances,

- qualitative and quantitative information about constant and proportional systematic deviations as well as their causes and elimination,
- temporal stability (perishability) and storage information for samples, reagents, standard solutions, and reference materials,
- special notes:
  - probable limits on the applicability of the process,
  - special procedures such as readjustment, recalibration, use of internal standards (including evaluation),
  - determination and use of a blank,
- notes about disposal of reagents and occupational safety,
- literature references with further information (e.g., examples of applications).

The documentation is to be supplemented at a later time by means of interlaboratory test data, the routine analyses to follow, and their respective quality assurance processes (see Phase IV).

2
Phase II: An Analytical Process Becomes Routine; Preparative Quality Assurance

2.1
Introduction

Statistically supported quality assurance measures are used to prepare for the introduction of an analytical procedure into routine analysis, as well as to provide accompanying validation. In this way, it is guaranteed that only valid routine analytical procedures are used [2, 7, 27, 92, 114, 137, 196, 197, 219].

2.1.1
Objective of Phase II

The objective of Phase II is to make a verified (see Phase I) analytical process available for routine analysis. Above all, this means that a sufficient analysis quality must be obtained and maintained *before* routine analysis may begin.

2.1.2
Execution of Phase II

In contrast to Phase I, which is carried out primarily by someone who is establishing a new analysis, Phase II is to be performed before the first routine use of each analytical process. Phase II is primarily of concern to the personnel who will be performing the routine analyses.

2.1.3
Progression of Phase II

Phase II begins with a "training" phase of the analytical process and ends with the preparation of quality control charts for the routine use to follow (see Scheme 2-1).

Scheme 2-1 Progression of Phase II.

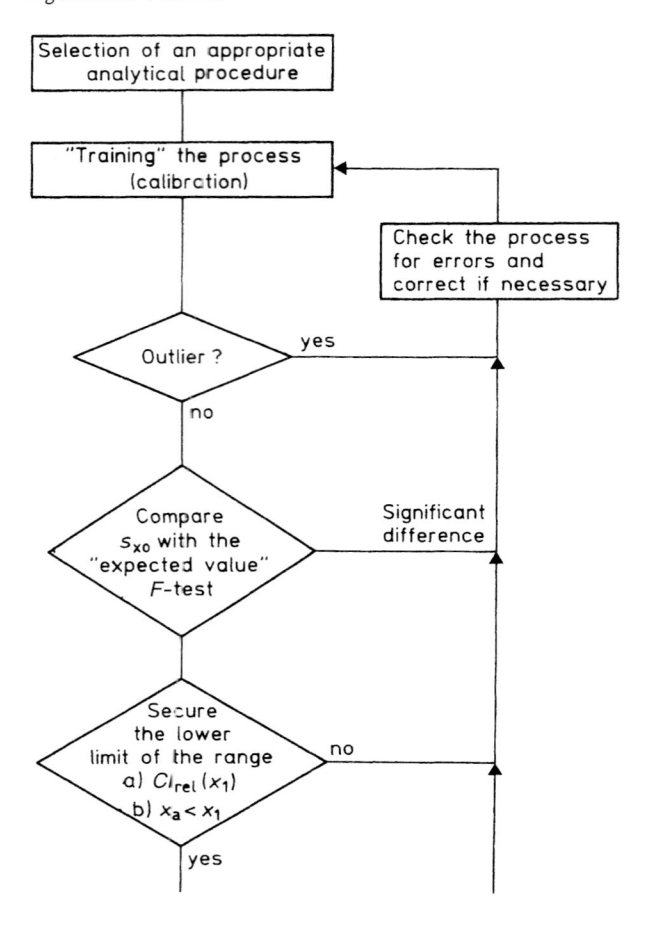

```
           ┌─────────────────────────┐
           │ Selection of an         │
           │ appropriate             │
           │ analytical procedure    │
           └───────────┬─────────────┘
                       │
           ┌───────────▼─────────────┐
           │ "Training" the process  │◄──────┐
           │      (calibration)      │       │
           └───────────┬─────────────┘       │
                       │                     │
                       │       ┌─────────────────────────┐
                       │       │ Check the process       │
                       │       │ for errors and          │
                       │       │ correct if necessary    │
                       │       └─────────────────────────┘
                       │                     ▲
                       ▼          yes        │
                    ◇ Outlier? ─────────────►│
                       │                     │
                       │ no                  │
                       ▼                     │
                    ◇ Compare                │
                      S_xo with the    Significant
                      "expected value"  difference
                      F-test ──────────────►│
                       │                     │
                       ▼                     │
                    ◇ Secure                 │
                      the lower              │
                      limit of the range  no │
                      a) CI_rel(x_1) ───────►│
                      b) x_a < x_1           │
                       │
                       │ yes
                       ▼
```

2.1.4
Results of Phase II

At the end of Phase II, the laboratory will not only have gained sufficient confidence and experience in the application of the analytical process in question, but it will also possess precision and reliability data which will serve during the routine period as a basis for quality assurance and control. This includes precision data such as the mean and the standard deviation of the analyses from control samples, the range of multiple analyses, the recovery rates of control samples, and especially the limits of reliability (scatter ranges, confidence intervals) within which future control analyses, or data derived therefrom, must lie.

```
        │
┌───────────────────────────┐
│ Establishment of quality  │
│ objectives                │
│ a) external quality requirements │
│ b) internal quality requirements │
└───────────┬───────────────┘
            │
┌───────────────────────────┐
│ Selection of a control chart │
│ system and suitable       │
│ control samples           │
└───────────┬───────────────┘
            │
┌───────────────────────────┐
│ Preparation of routine    │
│ quality assurance =       │
│ preliminary period        │
└───────────┬───────────────┘
```

Variance analysis $s_b \leq 2s_w$? — no →

↓ yes

Does the RR meet quality objectives? — no →

↓ yes

End of phase II

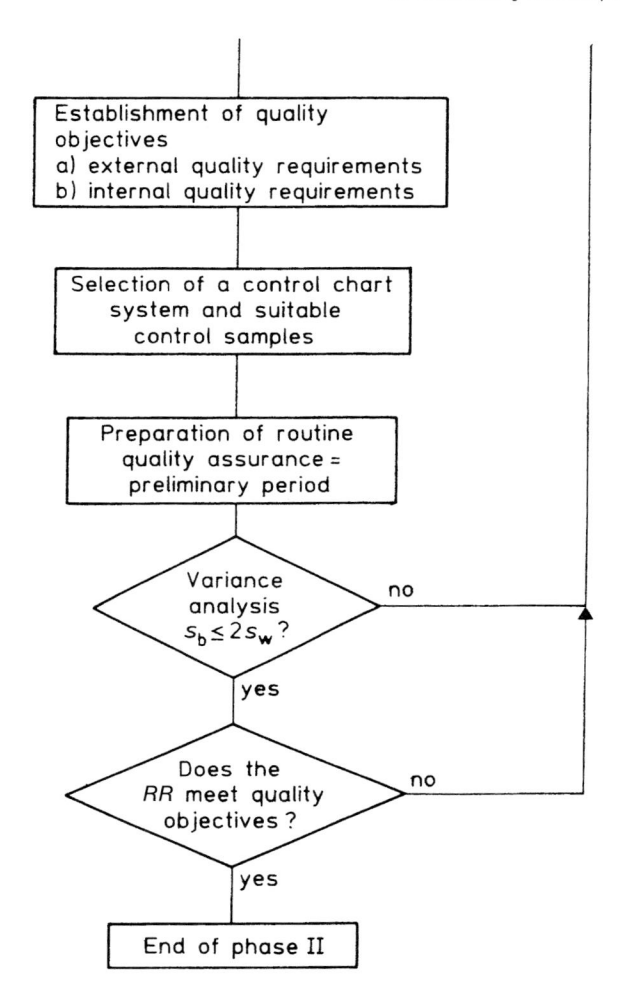

2.2
Selection of the Analytical Procedure

The choice of an analytical procedure is determined by the specific aspects of the analytical question posed, or in other words by:

- the substance to be determined:
  - which chemical element or compound,
  - the physicochemical state in which it occurs (dissolved, suspended, gaseous; oxidation state, etc.),
- the working range of interest,
- the required accuracy (maximum acceptable systematic and random deviations),
- the matrices to be investigated,
- the analytical equipment available, etc.

In addition, economic requirements (e.g., analysis time, workload and demands on staff, environmental and safety concerns, costs, etc.) must be taken into account. These should, however, under no circumstances be at the expense of quality.

2.2.1
Specificity of the Procedure

In routine analysis, only procedures that are able to determine all relevant forms of the substance, and are therefore specific, should be used. Many substances can exist in various physical and chemical forms (e.g., iron compounds are found in soluble, colloidal, or particle form; in different oxidation states; in complexes). It must be confirmed that the analytical procedure chosen covers all required forms [17, 28].

2.2.2
Selectivity of the Analytical Procedure

Additionally, one must be certain that the simultaneous presence of other substances in the matrix does not lead to incorrect results. Substances that cause errors (primarily systematic) in the determination of the analyte are known as *interfering substances*.

Interfering substances may be identified by means of *interference tests*, the experimental performance of which is described in Section 1.4.2.2.

2.2.3
Working Range

The working range of the chosen analytical procedure must be adapted to the problem. Dilution or concentration steps which are not described in the protocol should only be introduced into the process after testing (see Section 1.4.3.1), if at all, since they can increase the inaccuracy of the entire process.

2.2.4
Calibration Function, Sensitivity, and Precision of the Procedure

The calibration function determined by the process developer (see Section 1.2.3) produces results related to the sensitivity (slope, b), precision (residual standard deviation, s_y), and performance (process standard deviation, s_{xo}) of the analytical procedure.

The user of the process performs the calibration as described in the protocol in his or her own laboratory and compares the data obtained, especially the process standard deviation s_{xo}, with the data provided.

2.2.5
Minimum Detectable Value and Limit of Quantification

The minimum detectable value, x_{MDV}, of an analytical process is especially of interest if a substance is to be analyzed at the trace or ultra-trace level. In such situations, the analyst must choose the process with the greatest sensitivity and lowest possible imprecision, that is, the one with the lowest minimum detectable value.

One should keep in mind that an analytical result which corresponds to the minimum detectable value, x_{MDV}, indicates a relative confidence interval of 50%.

The minimum detectable value of the fundamental analytical procedure (see Section 1.3.2) should not be the sole consideration since this "ideal minimum detectable value" is obtained by analyzing pure standard solutions that have not been subjected to digestion, extraction, etc. The minimum detectable value for actual matrices can be considerably greater (by a factor of 5 to 10, based on experience). The reason for this is a possible negative influence of matrix components on the residual standard deviation, s_y, and the slope, b, of the calibration function. Therefore, one must not only consider the results of the process developer with respect to matrix influences, but also the "actual minimum detectable value" for one's own typical matrices using the method of standard addition (see Section 3.4.1.1).

2.2.6
Risk of Systematic Error

Since the process developer cannot determine a recovery function for all possible types of matrices, the user should perform additional recovery experiments (see Section 1.4) with his or her own matrices before beginning to use an analytical process in routine analysis.

2.2.7
Effort, Costs

Another criterion to be considered in the choice of an analytical process is the cost per analysis.

The following component factors should be evaluated with respect to their cost-effectiveness:

- *Time required*: This includes all the steps of sample preparation and analysis and can be determined from the analytical protocol.
  In addition to the time required for the analysis of actual samples, the following procedures also involve time, which must be included in the overall determination:
  a) quality assurance measures:
     – adjustment*
     – calibration*

- training of personnel*
- control analysis within the framework of internal *and* external quality assurance*
- maintenance* including inspections and preventive maintenance of equipment and care of the laboratory and equipment

b) evaluation and administrative work.

All items marked with an asterisk* are nearly independent of the number of samples in an analysis series and can be termed "fundamental requirements" in that they can influence the material requirements significantly.

Rationalization and, in particular, automatization measures are possible if the frequency of an analysis is high, and can significantly lower the average time required per analysis.

- *Necessary equipment*: Acquisition of equipment for a new analytical process incurs investment and operating costs, which are only economically justifiable at a higher analysis frequency.

- *Material required*: Analytical processes can differ markedly in their consumption of materials in relation to quality, quantity, and ease of storage.

- *Personnel required*: The number of persons and their qualifications must be considered and determined.

- *Type of application*: The effort involved in a process may vary depending on whether the process is used solely for qualitative analysis or if it can also be used for quantitative analysis, and if it is suitable for analyzing different types of matrices or if it possesses a very limited range of usage. The same is true for the evaluation of the working range, which must not only be appropriate for practical use but should be as large as possible in order to minimize calibration and adjustment effort.

In summary, the *frequency of usage* should be estimated for an analytical process. Processes which are seldom used can prove to be very cost intensive when the required quality assurance measures represent a large portion of the total costs. However, abandoning quality assurance measures on the basis of cost is unacceptable in all cases.

2.3
The "Training" Phase of the Process

The so-called "training" phase includes the determination of the quality required as well as the testing and, if necessary, the improvement of the quality achieved by the analytical process. Problems, which may appear in the analytical process itself, are to be investigated. The calibration of the analytical process follows this phase.

The process standard deviation, s_{xo} (see Section 1.2.3.1), is a measure of the precision of an analytical process. It is published in the process documentation within the framework of the characteristic data and represents the first quality objective to be obtained in the practical introduction of an analytical process. The appropriate calibration function is obtained during the "training" phase of the analytical process and it is used to calculate the process standard deviation.

The resulting process standard deviation is compared to the value stated in the process documentation by means of an F-test. If the expected process standard deviation has not been obtained, then the precision of the analysis must be improved. One additional "training" phase is almost always sufficient. Calibration and verification of precision must also be repeated according to the F-test.

If no significant difference can be found (see example, Table 2-1), then it may be assumed that the user has mastered the process.

Example:

"Training" phase of an iron determination (Table 2-1).
Desired value: s_{xo} = 0.034 mg/l, value from the table $F(f_1, f_2, P = 99\%) = 6.03$.

Table 2-1 Verification of the process standard deviation using the F-test.

Day	Process standard deviation mg/l Fe	Test value	Decision
First	s_{xo1} = 0.098	8.31	significant difference (from desired value)
Second	s_{xo2} = 0.087	6.55	significant difference
Third	s_{xo3} = 0.053	2.43	no significant difference
Fourth	s_{xo4} = 0.038	1.25	no significant difference

After obtaining the desired process standard deviation from the second repetition (third day), the user documents an acceptable precision as compared to the required analytical quality. If it is possible to maintain this target precision over a long time period, then this also describes the reliability of the analytical process. In other words, the level of performance of the pure analytical process should be maintainable in the future.

However, if a significantly higher imprecision (s_{xo} value) remains after numerous repetitions of the calibration, then the causes of this must be found. Some possible causes are:

- insufficient quality of the reagents used,
- standard solutions were not prepared accurately,
- misunderstanding of, or deviation from, the analysis protocol,
- use of inappropriate auxiliary equipment such as pipettes (see Table 2-2) or sample containers,
- use of inappropriate, unvalidated, or defective measuring apparatus.

Example: Optimization of a pipetting step

The calibration for iron described above was performed a total of four times, each using a different type of pipette for the preparation of the standard solutions. The results of the calibrations can be seen in Table 2-2:

Table 2-2 Optimization of a pipetting step.

Preparation of standard solution using	Slope b $\frac{abs.}{mg/l}$	s_y abs.	s_{x0} mg/l
1. Disposable pipette	0.999	0.0037	0.0038
2. Volumetric pipette	1.044	0.0011	0.0011
3. Microburette	1.059	0.0013	0.0012
4. Grease-free microburette	1.050	0.0005	0.0005

Additional error possibilities are described in Section 3.5.1.

If the lower limit (x_1) of the chosen range is lower than is described in the analytical protocol, then a smaller process standard deviation can often be obtained. However, there is a higher risk of obtaining uncertain analytical results at lower substance concentrations. Therefore, the relative imprecision at the lower end of the range must be checked:

a) if $CI_{rel}(x_1)$ exceeds the maximum allowable imprecision, then the analytical precision is in need of additional improvement;
b) if no minimum precision (maximum allowable imprecision) is defined, then the range is at least secured based on the test value x_a (see Section 1.2.4.2.3).

2.4
Establishment of Quality Objectives to be Adhered to in Routine Usage

The purpose of analysis is to produce reliable and comparable analytical results, not only in the short term but also in the long term.

The desired precision of the analytical results can only be achieved if the analyst is aware of the errors that may arise during the entire analytical process and is able to eliminate them. Only continuous monitoring and documentation of the analytical quality can guarantee and, if necessary, prove the precision of the analytical results.

The objective of a routine quality check is the rapid recognition of imprecision and errors in the analytical results with justifiable expenditure of effort.

One possibility for achieving this is laboratory-internal statistical quality control, which has long been used in the manufacturing industry for the control of product quality. It allows the identification of random errors through a *precision check* and of systematic errors through a *trueness check*. The quality targets of maximum allowable imprecision and inaccuracy, defined as absolute or relative errors, form the basis of quality control. Adherence to these targets during routine analysis is monitored through the systematic repetition of control analyses. Instruments of statistical quality control are quality control charts, which consist of consecutive visualizations of actually achieved quality (graphical presentations of the results of the control analyses). Recorded limit lines make it possible to control the quality:

- Control limits, transgression of which necessitates immediate corrective intervention in the analytical process. These limits are also known as action limits.
- Warning limits, transgression of which warrants special attention to the subsequent control analyses.

The position of the limits in the diagram is determined exclusively by precision quantities (standard deviation, etc.).

When establishing a quality control chart, the extent of scatter of the analytical results should be critically evaluated. The standard deviation, and with it the warning and control limits, may not exceed acceptable magnitudes or legal requirements. For example, it would not make sense to use a Shewhart chart to monitor an analytical process which has a variation coefficient of 50%.

In many areas of analysis, for example drinking water analysis, the precision and trueness requirements are partly set by national legislation [186, 205]. If such standards are not present, then acceptable limits of error can be determined as follows:

a) Use of the precision data from the standard process method as guidelines.
b) By comparison with other laboratories. Whether as part of a round-robin test or through personal exchange, quantities obtained in one's own laboratory should be checked before they are transferred to a control chart. In addition, the literature should be consulted on this matter.
c) Adherance to legal minimum requirements (for example, European Community guidelines).
d) Other authors (e.g., Tonks [200]) recommend that the acceptable relative limits of error should be calculated from the position and width of the range.

2.4.1
External Quality Requirements

Environmentally relevant laws, decrees, guidelines, regulations, and international agreements increasingly require not only the observance of limits, but also a specific analytical quality, quantified in terms of either accuracy or precision. For example, an EU directive regarding the purity of drinking water [186], and the national regulations derived from it, or the guidelines for laboratory medicine [187, 188], indicate the acceptable error of analytical results for the individual parameters. Furthermore, the quality guidelines that need to be met can be stipulated in approved levels for emissions (i.e., effluent, exhausts).

Example: The sulfate content in drinking water should be monitored. The concentration limit and maximum acceptable analytical uncertainty are set as follows:

Concentration limit: 240 mg/l SO_4^{2-}
Acceptable error: ± 5 mg/l $SO_4^{2-} \cong 2.08\%$; i.e., $CI_{rel}(x = 240) = 2.08\%$

2.4.2
Internal Quality Requirements

If neither the client nor the governing bodies provide any information with respect to the required quality of an analysis for a specific parameter, or if the user requires an analytical process of higher quality, then the analyst may set his or her own quality targets. A minimum requirement is that the precision obtained in the training phase must be maintained during the subsequent routine analysis. If round-robin data (see Section 4.3) for the analytical process are known, then the published repeatability standard deviation, s_r, may be used as a precision target, provided that the substance contents and especially the matrices are similar.

2.5
Control Samples for Internal Quality Assurance

Since analytical quality needs to be monitored over a longer period of time through control analyses, so-called control samples, i.e., samples that *remain stable over a long period of time*, must be analyzed as controls in each analysis series.

2.5.1
Requirements of Control Samples

To be usable as a control, a sample must generally fulfill the following requirements [195]:

- The control sample should be representative with respect to its matrix and concentration.
- The content of the control sample should be chosen so that analytically important regions (e.g., the limit regions) can be secured.
- The control sample should be available in sufficient quantity, so that control analyses can be performed over a long period of time using the same sample.
- The stability of the control sample must be proven and should be maintained for at least several months (under defined storage conditions).
- There must be no influence of containers on the shelf-life of the control samples.

- The regular removal of sample aliquots for the control analyses must not lead to changes in the control sample (e.g., evaporation of highly volatile components through opening of the container).

2.5.2
Types of Control Samples

For quality assurance, the following may be used as control samples:
- standard solutions,
- blank samples,
- natural samples,
- synthetic samples,
- certified reference materials.

The characteristics of these various control samples, that is, their properties and peculiarities, as well as the demands placed on them, are discussed in the following subsections.

2.5.2.1 Standard Solution
A standard solution is a solution of a standard substance of known purity in a solvent as similar as possible to that of the sample.

In preparing standard solutions, the standard substance is weighed to a precision of 0.1%; including the volume errors of the solvent, the precision must be 1% or better. Standard substances and solvents themselves must be of sufficient purity.

The content of these solutions is therefore known to 1% uncertainty.

2.5.2.2 Blank Samples
A blank sample is an analysis sample that is presumed to be free of the component to be determined, but is submitted to the entire analysis process in the same way as an actual sample. The result of this analysis is known as the blank value.

2.5.2.3 Natural Samples
The natural sample, sometimes also misleadingly called the standard sample [195], contains many unspecified unknown components in addition to the analyte.

The content of these actual samples must be analytically determined beforehand. Since neither their true nor their conventionally true values are known, they may not be used as expected values for the accuracy check; actual samples are suitable only for precision checks.

2.5.2.4 Spiked Natural Samples

A natural sample, regardless of its original substance content, is used as a representative matrix for a recovery experiment. First, the original sample is analyzed before the addition of a standard. Then, a defined quantity of a standard substance or standard solution is added, the combination is mixed sufficiently, and the second analysis is performed. The control quantity is the recovery rate (RR), the quotient of the found difference in substance content and the expected difference in substance content, expressed in percent.

The addition of the standard substance/solution must not, however, lead to a volume error. Otherwise, a corrective calculation taking into account the modified volume must be carried out.

Example:

Volume of the original sample:	$V_o = 10$ ml
Result of analysis of the original sample:	$\hat{x} = 73$ µg/l
Added standard concentration:	$x_{add} = 75$ µg/l
Result of analysis of the spiked sample:	$\hat{x} = 151$ µg/l
Recovery rate:	$RR = \dfrac{151 - 73}{75} \cdot 100\% = 104\%$

If the volume error must be taken into account:

Added volume of the standard solution:	$V+ = 1$ ml
Corrected x_{add}:	$x_{add} = \dfrac{75\ \mu g/l \cdot 10\ ml}{10\ ml + 1\ ml} = 68.18\ \mu g/l$
Corrected recovery rate:	$RR = \dfrac{151 - 73}{68.18} \cdot 100\% = 114\%$

2.5.2.5 Synthetic Samples

A synthetic sample is a standard solution that, in addition to the analyte, contains other components (possibly interfering components) in known concentrations. The preparation precision is the same as for standard solutions (see Section 2.5.2.1).

2.5.2.6 Certified Reference Materials (CRMs)

Certified reference materials – sometimes also erroneously called standard samples – are natural or synthetic samples, the content of which is established by various persons by means of various analytical methods. Certified reference materials, along with their expected property values and the determination uncertainty, are offered by accredited certifying bodies (e.g., NIST, BCR). Due to the high degree of effort in determining the expected property values, they are very expensive.

2.5.3
Requirements for Producers of Control Materials

Commercially available control samples, the production of which is not the direct responsibility of the analyst who uses them, should only be accepted when furnished with additional information [131]:

1. A statement about the expected value and its measurement uncertainty.
2. The expiry date of the control sample for unopened sample containers under various defined storage conditions; the date after which the product should not be used.
3. The period of time over which a control sample may still be used after opening of the sample container, with distinction according to storage conditions (storage at room temperature, at +4 °C, at -18 °C).
4. Instructions for handling the control sample. For frozen samples: information as to whether, how often, and in what manner the sample may be thawed and refrozen.
5. Information about how the expected value ± uncertainty has been determined as well as suggestions for analysis procedures to be used.
6. A statement about possible additional components in the sample and their concentrations.
7. Information about the origin of the bulk (or gross) sample and a statement about the amounts and purities of components added to it.
8. Information regarding the homogeneity of the control material. For use as a precision control, a homogeneity of $\geq 99.5\%$ is required, verifiable by means of repeated analyses of aliquots from the same sample container (maximum variation coefficient $\leq 0.5\%$).

If a laboratory uses its own control samples, especially natural samples, then the same requirements apply.

2.5.4
Applicability of Control Sample Types (see Table 2-3)

Control samples should be co-analyzed at least once or twice in each analysis series.

While a precision control merely requires a homogeneous sample that is stable over a long period of time, the trueness of analytical results can only be definitively assured using a control sample with an expected value that has been sufficiently investigated.

Since matrix effects can have significant influence on the accuracy of the results, pure standard solutions, for example, are not suitable for practical trueness testing. A trueness control sample must be representative of the samples (matrices) that are to be tested routinely. If necessary, a quality assurance must be carried out for each individual problem matrix.

Table 2-3 Types of control samples and their internal quality assurance applications.

Sample type	Precision	Suitability for monitoring Trueness (limited)	Trueness
Standard solution	yes ($\bar{x}$-chart)	no	no
Blank sample	yes ($\bar{x}$-chart)	yes (monitoring of reagents)	no
Natural sample	yes (R-chart)	no	no
Natural sample + spiking	no	yes (recovery chart)	no
Synthetic sample	yes	yes (only if representative of the matrix)	yes (only if representative of the matrix)
Certified reference material	yes	yes	yes

($\bar{x}$ = mean value, R = range)

2.6
The Control Chart System

2.6.1
Introduction: History of the Control Chart

The quality control chart, developed by Shewhart in 1931 [193], was originally used for industrial product control. In contrast to laboratory analysis, the problem consists of a product which should possess constant features within certain stated limits. The length of a screw represents an example of such a feature. During the manufacturing process, any desired number of items can be spot-checked, i.e., the screws can be measured. When monitoring a characteristic, Shewhart took several random samples within a manufacturing step (e.g., day or machine-dependent period). These n samples then belonged to a subgroup. To prepare the control chart (expected value and establishment of warning and control limits), he then performed analyses on N subgroups. The property in question was also investigated routinely n times and the corresponding mean was entered into the control chart (see Table 2-4 and Figure 2-1).

Serious changes which suddenly appear in the production conditions, as well as a slow but steady decline in quality, can be read directly from the graph (for details, see Section 2.6.7.1). Immediate corrective intervention in the production process decreases the risk of "scrap production" and customer complaints.

Through these multiple measurements, Shewhart monitored not single values of a property, but mean values. This procedure could overwhelm all routine analysis in an analytical laboratory as, for example, control samples must be tested re-

2.6 The Control Chart System

Table 2-4 Table of original values for product control according to Shewhart and related control charts.

Measurement number j within the group	Production step i (= group)							
	1	2	3	4	.	.	.	N
$j = 1$	$x_{1,1}$	$x_{2,1}$	$x_{3,1}$	$x_{4,1}$				$x_{N,1}$
$j = 2$	$x_{1,2}$							
$j = 3$	$x_{1,3}$							
$j = 4$	$x_{1,4}$							
.	.							
.	.					x_{ij}		
.	.							
$j = n$	$x_{1,n}$							
Group mean	$\bar{x}_1$	$\bar{x}_2$						$\bar{x}_N$

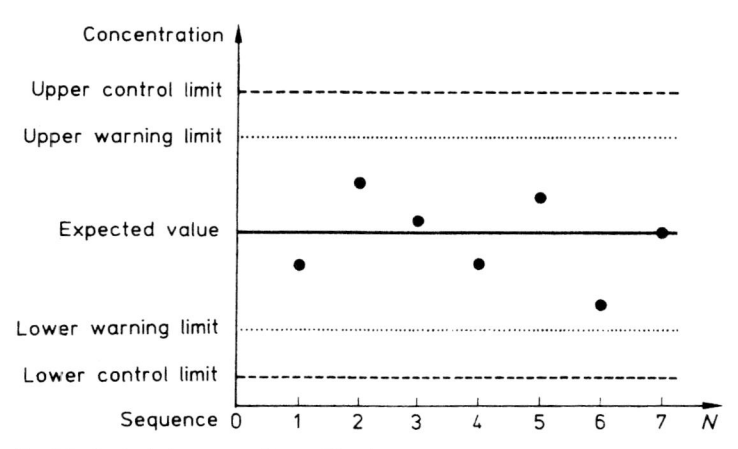

Fig. 2-1 Control chart according to Shewhart.

peatedly for every parameter in every series. Therefore, most of the subgroups described here consist of only one analysis result, i.e. $n = 1$.

Another fundamental difference to product control in industry is the necessity to produce one's own "control product" (namely control analyses). The actual "product" of an analytical result eludes a precision or accuracy check; at best only a plausibility test can be performed. In addition, the analysis result as "product" may, in an individual case, assume every value within the range. Equivalence with previous or successive results for various analysis samples would be pure coincidence.

2.6.2
Principle of a Control Chart

The principle of a control chart is a visual representation of "quality" based on:

- the desired quantity (expected value of the analysis results for control samples),
- the quality limits (see Figure 2-2).

For the latter, one distinguishes between

- *warning limits*, a single transgression of which can be tolerated, and
- *control (or action) limits*, transgression of which necessitates immediate action.

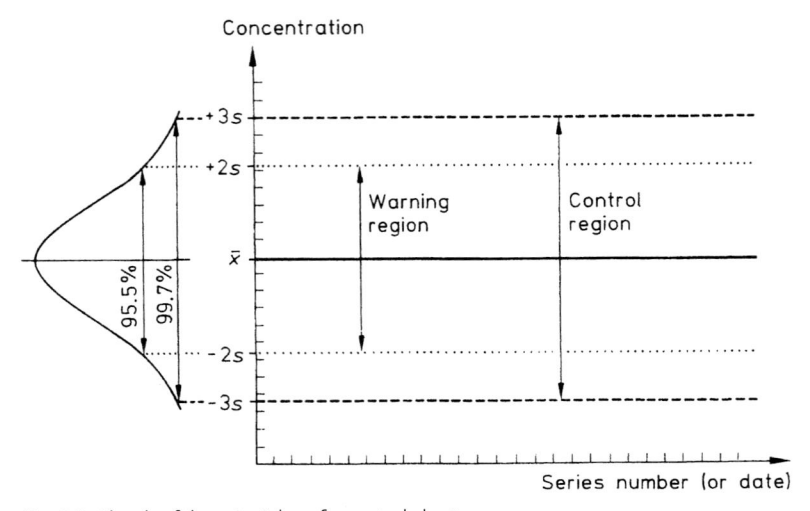

Fig. 2-2 Sketch of the principles of a control chart.

In a so-called preliminary period, one or more control samples are co-analyzed in every analysis series.

For each control sample, a precision and – as far as possible – an accuracy test is then performed on the analytical results (comparison with established values).

For acceptable analysis quality, the warning and control limits are established and the control charts are constructed.

In the following control period, the adherence to obtained limits is monitored and additional control samples are investigated.

The results of these control analyses, or the statistical data obtained from them, are continuously entered in the control chart. The abscissa represents the sequence (batch number or working day); results are entered on the ordinate. The progression of results is visible at once; the control chart can be visually evaluated after each additional entry.

2.6 The Control Chart System

The control chart thereby allows the rapid recognition of arising errors, and established control criteria indicate an "out-of-control situation" for the analytical process being monitored.

Quality assurance is a regulation process. Every analytical series and control sample must undergo the regulation cycle at least once:

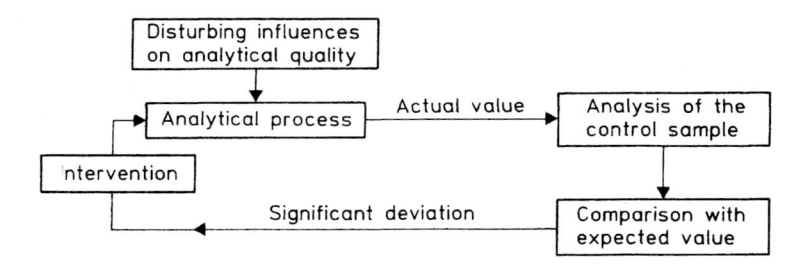

```
           ┌─────────────────────────┐
           │ Disturbing influences   │
           │ on analytical quality   │
           └──────────┬──────────────┘
                      ▼
    ┌──────────────────────┐  Actual value   ┌──────────────────────┐
 ┌─▶│ Analytical process   │────────────────▶│ Analysis of the      │
 │  └──────────────────────┘                 │ control sample       │
 │                                           └──────────┬───────────┘
 │  ┌──────────────┐                                    │
 │  │ Intervention │                                    │
 │  └──────┬───────┘                                    ▼
 │         │        Significant deviation   ┌──────────────────────┐
 └─────────┴◀──────────────────────────────│ Comparison with      │
                                            │ expected value       │
                                            └──────────────────────┘
```

Using a control chart, any one of the following statistical quality data parameters can be monitored:

- single value, mean value,
- recovery rate,
- standard deviation,
- range.

2.6.3
Average Run Length (ARL) and Evaluation of Control Charts

Limits for control charts are fixed from two perspectives:

1. Notification of a so-called statistically "out-of-control" situation as rapidly as possible, if the analytical process changes significantly (the measured values belong to another entity) due to unknown influences on the analytical process (or product, in industrial applications).

2. A small likelihood of a "false alarm", i.e., a measured value which actually belongs to the same entity (the chart), but lies outside the limits incidentally. When the chosen limits are too narrow, one must accept the possibility of frequent false alarms.

An important parameter for the evaluation of control charts and out-of-control situations is the average run length (ARL). The ARL is defined as the average number of entries on a control chart before a single out-of-control situation appears [109, 125, 216].

A high ARL (L_0) is necessary if the analytical process is to proceed in a controlled manner. The possibility of a "false alarm" (α-error) is then very small. According to Woodward [216], L_0 should be of the order of several hundreds. For the mean value chart, on average one value out of 400 falls outside the $3s$ limit, even though the procedure is under control. The ARL of the mean control chart is therefore $L_0 = 400$ [148].

On the other hand, an actual out-of-control situation should be indicated as quickly as possible, i.e. the ARL (L_1) must be very small. Haeckel [109] and Woodward [216] consider an L_1 between 3 and 10 as reasonable and obtainable with conventional control charts.

2.6.4
Derivation of the Average Run Length (ARL) [106]

Assuming a mean value control chart having only 3s control limits, an equation is derived for the ARL as follows:

$\bar{\bar{x}}$ \quad the grand average

$\bar{\bar{x}} + 3 \dfrac{s}{\sqrt{n}}$ \quad the upper control limit

(In this case, n analyses per subgroup are performed, the mean $\bar{x}_i$ is then entered on the chart.)

The probability that a subgroup mean lies outside of the control limits is only 0.003, or 0.3%. This value is obtained from the Gaussian curve, see Figure 2-2. On average, three per thousand, or one out of every 333 values, lies outside of the control limits. If the grand average is shifted by K standard deviations according to

$\bar{\bar{x}} + K \cdot s$

the upper (old) control limit lies at a distance of

$$e = \bar{\bar{x}} + \frac{3s}{\sqrt{n}} - (\bar{\bar{x}} + K \cdot s) = s\left(\frac{3}{\sqrt{n}} - K\right) \qquad (64)$$

from the new grand average value (see Figure 2-3).

This distance (e) is transformed into the standardized normal distribution unit e' by dividing it by $s/\sqrt{n}$:

$$e' = 3 - K\sqrt{n} \qquad (65)$$

The probability that a subgroup mean lies outside of the (old) upper control limit after shifting the grand average by $K \cdot s$ is the same as the probability of obtaining a value larger than the standardized distance, e'. This is the probability that a change in the grand average by $K \cdot s$ will be detected (sensitivity of the chart). Since this probability should of course be very large, it is evident that the distance e' needs to be minimized, for example by increasing the number of analyses, n, performed per subgroup.

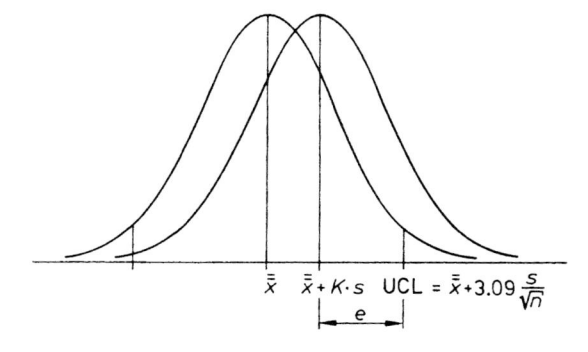

Fig. 2-3 Shifting of the grand average by $K \cdot s$.

2.6.4.1 Examples of Theoretical Calculations

a) The number of analyses per subgroup is $n = 1$ and the grand average has increased by $1s$ ($K = 1$). For $n = 1$ and $K = 1$, the value 2 is obtained using Eq. (65):

$$e' = 3 - 1 = 2$$

The probability that a value greater than e' results from an analysis ($n = 1$) can be obtained from a standard normal distribution table [127].
For $e' = 2$, the resulting probability is $1 - 0.9772 = 0.0228 = 2.28\%$.

b) If the subgroup quantity (for unchanged $K = 1$) is $n = 4$, then the resulting distance $e' = 1$.

$$e' = 3 - 2 = 1$$

The probability that a value is larger than 1 is 15.87% (0.1587).

The sensitivity increases with increasing number of measurements per subgroup.

The sensitivity is quantified through the ARL:

$$\text{ARL} = \frac{1}{\text{probability of the appearance of a value larger than } e' \text{ for a change in the grand average of } K \cdot s}$$

Figure 2-4 depicts the progression of ARL curves for three mean value control charts with varying subgroup sizes n (single out-of-control criterion: $3.09 \times s$ control limits). The x-axis indicates the value of the systematic deviation of the mean value in question from the original (expected) mean value (expressed in standard deviation units). A deviation of $0 \cdot s$ indicates that no systematic errors are present; consequently, one out of 500 mean values will randomly lie outside of the control limits ($L_0 = 500$).

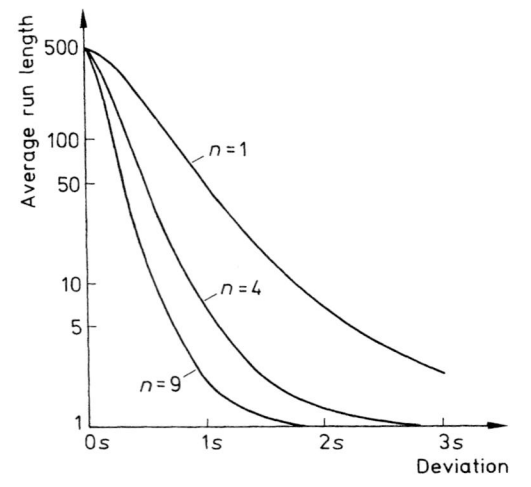

Fig. 2-4 ARL curves for three $\bar{x}$-charts with varying number of replicates n.

For only one control analysis per series, an average of 50 series is required before a systematic error in the quantity $1s$ is detected. For nine control analyses per series, however, the alarm is raised after just two to three series.

2.6.4.2 Analytical Example [141]

Mean $\bar{x} = 1000$ ppb, $s = 50$ ppb, $3 \cdot s = 150$ ppb

The control limits are therefore:

upper control limit, UCL = $\bar{x} + 3 \cdot s = 1150$ ppb
lower control limit, LCL = $\bar{x} - 3 \cdot s = 850$ ppb

An increase in the mean of 100 ppb (= $2 \cdot s$) to $\bar{x} = 1100$ ppb changes the $3 \cdot s$ range to 1250 to 950 ppb. However, the probability that a single value lies outside of the limits for the aforementioned increase in the mean is 16%. An average of 27 single measurements would be necessary to detect a change of this sort in the system with an error probability of 1%. By contrast, a mean value control chart obtained using means from four measurements would result in substantially narrower control limits:

$$\bar{x} \pm \frac{3s}{\sqrt{n}} = \bar{x} \pm \frac{3 \cdot 50}{\sqrt{4}} = \bar{x} \pm 75 \text{ ppb}$$

UCL = 1075 ppb
LCL = 925 ppb

2.6 The Control Chart System | 77

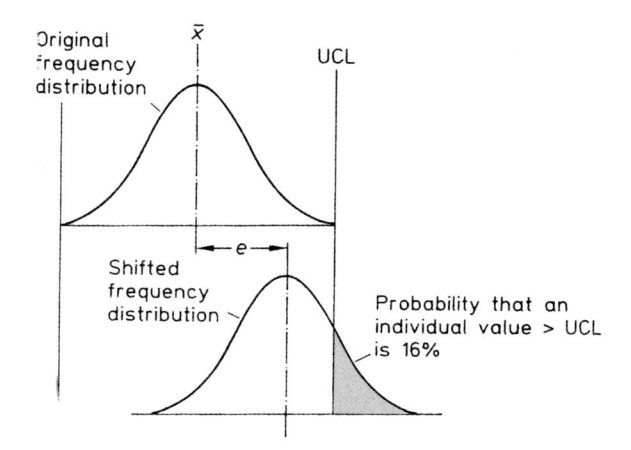

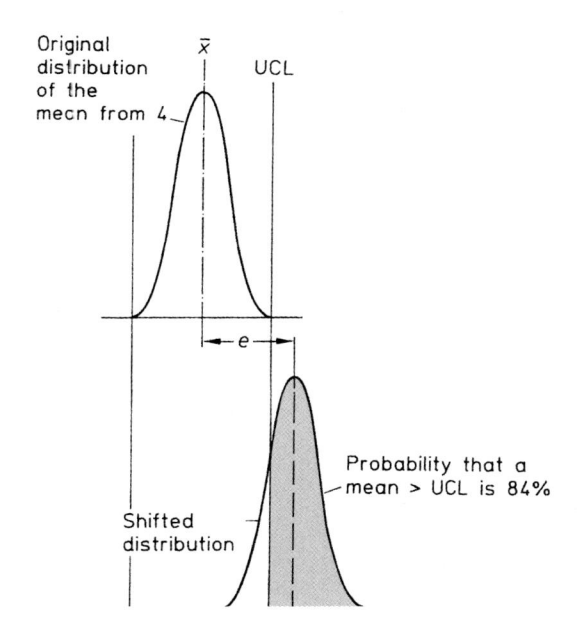

Fig. 2-5 Representation of the more sensitive mean value shift for quadruplicate measurements as compared to single measurements.

For a change in the grand average value from 1000 to 1100 ppb, the probability that a mean value from four measurements lies outside of the old limits is 84 out of 100 (cases). To determine the change with a probability of 99%, two or three subgroups from each of four measurements are sufficient (see Figure 2-5).

It is possible to combine the advantages of single values (which are less time-intensive) with those of mean values (which produce more sensitive charts) in the use of "moving ranges" (for an explanation, see Section 2.6.7.2).

2.6.5
Concept for the Preparation of Routine Quality Control (So-Called Preliminary Period)

Objectives of the preliminary period are:

- The collection of characteristic data to prepare the control chart.
- The preliminary reliability testing of the analytical process. For the preliminary period, a sequence of 20 consecutive working days [117, 139, 141, 211] or analysis series is designated, each involving duplicate determinations for at least the control samples listed below:
  1. blank samples,
  2. standard solutions each having one low and one high concentration within the established range,
  3. natural samples of typical concentration and matrix, and
  4. spiked natural samples.

In order to register changes/trends within a series of analyses on a given day, the two analyses of a duplicate determination are performed with as much time between them as possible.

These analyses are to be performed with the same care that a normal sample would receive. Special handling of the control sample during the preliminary period or routine analysis would falsify the message of the control chart.

Note: A preliminary period of 20 working days or analysis series results in sufficient statistical safeguards, both for the variance analysis and the establishment of warning and control limits. Whitehead [211] justifies the number of $N = 20$ analyses in the preliminary period by the fact that for $N > 20$ repeated analyses, the standard deviation of the mean decreases only slightly (see Figure 2-6). A shortening of the preliminary period to ten or even six series (the absolute minimum according to WHO recommendations [212]) is only acceptable if the control limits can be verified by other, external means.

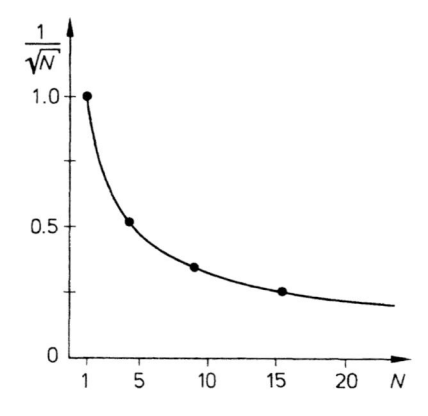

Fig. 2-6 Influence of the number of repeated measurements (N) on the size of the standard deviation ($s/\sqrt{N}$) of the mean [211].

2.6 The Control Chart System

Considering the four control sample types listed above:

1. Blank samples:
If necessary, determination of blank values can be used to determine the decision limit and minimum detectable value (see Sections 1.3.1 and 1.3.2).

More important, however, is the use of blank values for quality assurance:

a) Changes in reagents, new batches of reagents, carryover errors, drift of apparatus parameters, etc., can show up as a change in the blank sample.
b) Blank value determinations at the beginning and end of an analysis series allow the identification of systematic trends.

2. Standard solutions:
The primary objective of quality assurance by means of standard solutions is the verification of the calibration and the calibration conditions.

It is important that the standard solution to be analyzed is not identical to one of the calibration standard solutions, otherwise an error in the preparation of the standard solutions could not be detected. The results for the standard solution can be documented in a mean value control chart (see Section 2.6.7.1.5.1).

3. Natural samples:
If samples of very different composition (different matrices, such as waste water, landfill seepage water, surface water, biological samples such as body fluids, soil, air or dust samples) are to be analyzed routinely, then an actual sample of each matrix type should be selected, if possible.

For duplicate determinations of the content of these actual samples, the difference in the two results can be documented in a different R-chart (range charts, see Section 2.6.7.2) for each individual type of matrix. In this way, matrix-dependent disruptions of the precision can quickly be recognized.

4. Spiked natural samples:
If the result $\hat{x}$ is determined for a partial quantity of the actual sample, and another portion of the original sample is spiked with a concentrated standard solution (avoiding volumetric errors) and also analyzed, then the recovery rate can be determined from the difference in the two results:

$$RR = \frac{\hat{x}_S - \hat{x}}{x_{exp}} \cdot 100\,(\%)$$

$\hat{x}_S$ = analysis result for the spiked sample
x_{exp} = expected result due to spiking

A recovery rate (RR) control chart can be established for recovery rates obtained in this way (see Section 2.6.7.1.5.3), and it can be used to detect matrix disturbances, among other things. Each matrix type must have its own RR control chart.

It must be kept in mind that the type of substance used to spike the sample needs to be representative. For example, it does not make sense to determine recovery rates with potassium hydrogen phthalate (KHP) for the determination of chemical oxygen demand (COD) in water analysis, since KHP is very easily oxidized and ultimately only the pipetting error for the addition of the KHP solution will be recorded and not errors in the COD analysis or matrix-dependent disturbances.

2.6.6
Evaluation of the Preliminary Period

Even though the data from the preliminary period serve primarily for the establishment of control charts (see below) and may be recorded in these for checking purposes, they are also subjected to preliminary testing.

2.6.6.1 Variance Analysis

A variance analysis (see Section 1.7.1.1) is performed for each type of control sample (except the actual sample, which changes daily) after the end of the preliminary period. The standard deviation between batches (s_b) should not be more than twice the standard deviation within batches (s_w).

2.6.6.2 Adherence to Required Quality Objectives

The control limits determined during the preliminary period should be checked against the maximum permissible result uncertainty, if it is defined (see Sections 2.4.1 and 2.4.2). If the control limits exceed the specified maximum uncertainty, the analytical procedure must be optimized and the preliminary period extended.

2.6.7
Types of Control Charts and Their Applications

The following types of control charts are used in analytical quality assurance:

- Shewhart (or conventional) charts as
  - mean value control charts (also blank value and target value charts),
  - recovery rate charts,
- R-charts and standard deviation charts,
- cusum charts,
- target value charts.

Each type of control chart is described individually below.

2.6.7.1 Shewhart Charts

2.6.7.1.1 General

One of the oldest and simplest types of control chart is the Shewhart chart. It can be used to monitor the precision of mean values from multiple analyses or of single results [81, 124, 208]. Whether or not a newly determined value belongs to the pool of values necessary for the preparation of the chart is evaluated visually and objectively using a graph (i.e., results lie within or outside of limits or lines) [17]. Application of the chart described here requires a normal distribution and values with constant standard deviation.

2.6.7.1.2 Statistical Fundamentals of the Shewhart Chart

If the range obtained by repeated measurements of a sample is divided into classes of uniform size, and the frequency with which the individual results fall into a class is counted, then a histogram can be plotted (see Figure 2-7).

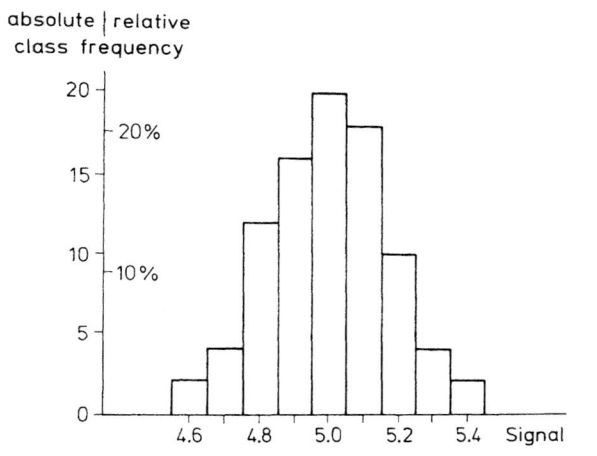

Fig. 2-7 Histogram (for small N) [109].

The number of classes is related to the number of individual results (N): number of classes = $\sqrt{N}$.

For very frequent repetition of the measurement ($N \to \infty$) and simultaneous infinite reduction of the width of the individual classes, the histogram approaches the form of a so-called Gaussian (or bell) curve. This normal distribution is described by the mean value of the individual results, as well as by the standard deviation. The mean of the measurements lies under the peak of the bell curve. The standard deviation is obtained from the intersection of a vertical line drawn through the point of inflection of the curve with the x-axis. The area between the two $1 \cdot s$ lines covers 68.3% of the entire area underneath the bell curve and therefore includes 68.3% of all individual results; the area between the $2 \cdot s$ lines covers 95.5% of the total and the area between the $3 \cdot s$ lines 99.7% (see Figure 2-8).

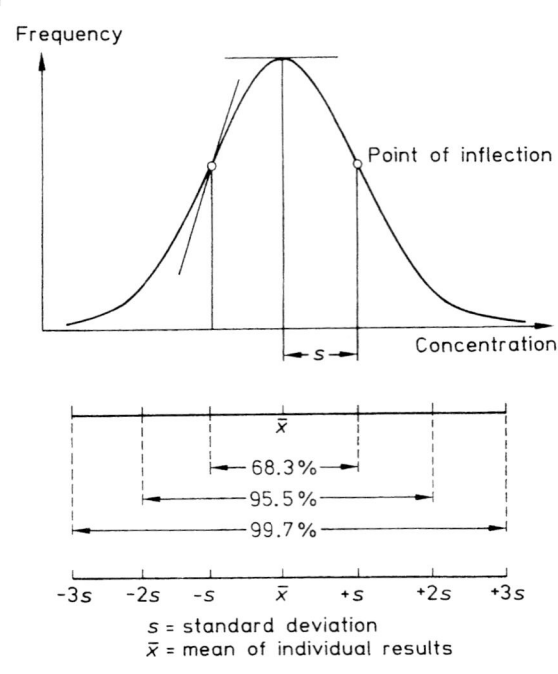

Fig. 2-8 Frequency distribution of the analysis results at $N \to \infty$ repeated analyses and class width $\to 0$ (Gaussian bell curve).

s = standard deviation
$\bar{x}$ = mean of individual results

The following conclusions can be drawn from this representative form:

- The entire area between curve and x-axis is always the same (= 100%), irrespective of $\bar{x}$ and s, and ends only in infinity.
- If the results are widely distributed around the mean, i.e. for very large standard deviation, then for the same total area this is represented by a wide but low bell curve.
- Conversely, a narrow, tall curve represents a small standard deviation, i.e., a high degree of precision.

2.6.7.1.3 Construction of a Shewhart Chart

The application of control charts in quality assurance is based on the assumption that the obtained results are normally distributed [3, 214]. Shewhart [193] therefore developed a control chart which features a bell curve rotated through 90°.

First, a system of coordinates is set up, in which the unit of the results is ascribed to the ordinate (y-axis) and is divided correspondingly (e.g., concentration units). The abscissa (x-axis) is a discrete time axis, on which the time, series number, or day, for example, is plotted in chronological form.

The results of the preliminary period are then entered consecutively.

After N days or series, the mean, $\bar{x}$, and the standard deviation, s, are calculated and the Shewhart chart is constructed. The central line of the chart is defined by the calculated mean value from the preliminary period (see Figure 2-9).

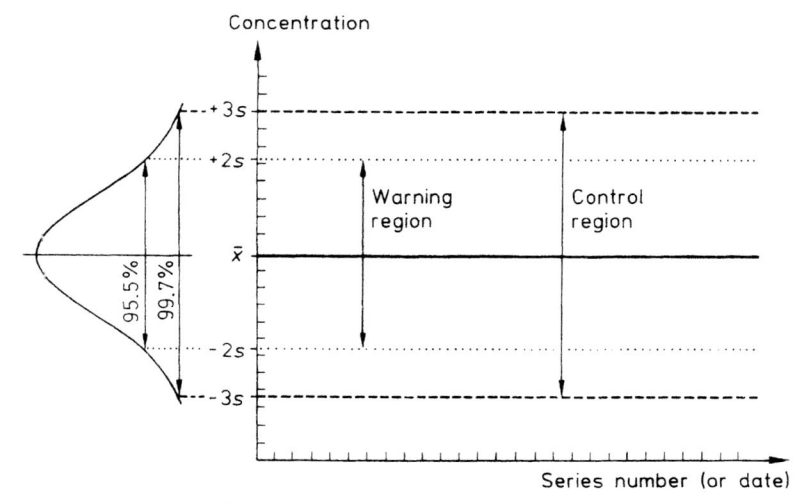

Fig. 2-9 Construction of a control chart.

With the aid of the standard deviation, the warning and control limits are drawn at $\bar{x} \pm 2 \cdot s$ and $\bar{x} \pm 3 \cdot s$, respectively.

A hint for construction is that the y-axis should always approximately cover the range $\pm 4s$. In this way, it is certain that space will be available to plot "out-of-control" values. The results of the preliminary period are then plotted on the prepared chart and the chart is checked according to the control criteria described in Section 2.6.7.1.4.1.

The control and warning limits are the $\bar{x} \pm 3 \cdot s$ and $\bar{x} \pm 2 \cdot s$ ranges, respectively:

- upper control limit, UCL = $\bar{x} + 3 \cdot s$
- lower control limit, LCL = $\bar{x} - 3 \cdot s$
- upper warning limit, UWL = $\bar{x} + 2 \cdot s$
- lower warning limit, LWL = $\bar{x} - 2 \cdot s$

The range of $\pm 2 \cdot s$ either side of the central line covers 95.5% of the area underneath the curve, i.e. the probability of a "false alarm" in this area is 4.5%, and a single transgression of this limit is tolerated. The probability of a value exceeding the $3 \cdot s$ limit is 0.3%, i.e., if this occurs, then it is with fair certainty an out-of-control situation.

During the evaluation of data from the preliminary period, the detection of an out-of-control situation already present in this period indicates that corrective measures are urgently required before routine analysis can begin (see Section 3.5).

2.6.7.1.4 Evaluation of a Control Chart

The more precise the analytical process is, the more sensitively a control chart reacts, i.e. recognizes out-of-control situations [1].

2.6.7.1.4.1 "Out-of-Control" Situations

In addition to the detection of large random errors (gross errors), the control chart should also provide indications of systematic errors or trends in systematic errors.

The following criteria for out-of-control situations are mentioned in the literature:

1. One value outside of the control limits [29, 56, 109, 117, 179] (no. 1 in Figure 2-10).
2. Seven consecutive values on one side of the central line [56, 109, 117] (no. 2).
3. Seven consecutive values showing an ascending trend [56, 109, 117] (no. 3).
4. Seven consecutive values showing a descending trend [56, 109, 117] (no. 4).
5. Two of three consecutive values outside of the warning limits [56, 141].
6. Ten of eleven consecutive values on one side of the central line [117].

Out-of-control situations 1 to 4 are depicted in Figure 2-10.

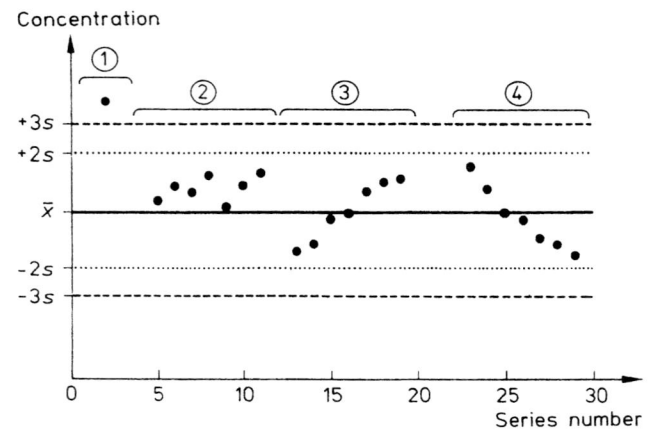

Fig. 2-10 Examples of out-of-control situations.

2.6.7.1.4.2 Conspicuous Entries

When evaluating a control chart, one should not only look for out-of-control situations, but should also follow the general progression of the entries on the chart. Figure 2-11 depicts four examples in which at no time is the process "out of control", but the order of entries suggests influences that are not random.

Action should be taken before the appearance of an out-of-control situation; in cases b), c), and d), out-of-control situations can be expected to arise in the foreseeable future.

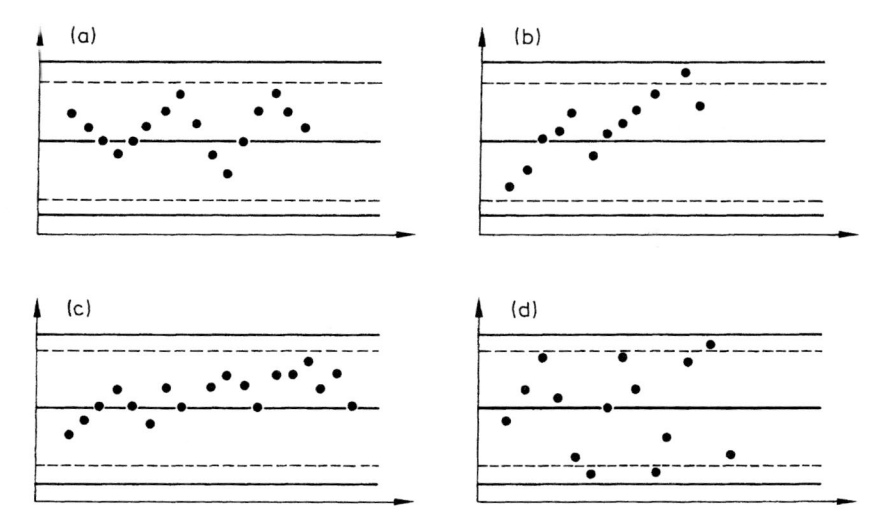

Fig. 2-11 Conspicuous order of entries in a Shewhart chart.
a) cyclical changes (cause: rotation of technician, "Monday" effect, etc.) [141];
b) shift of the mean (cause: technical intervention on the measurement equipment, new reagents, new equipment, disposable articles, etc.);
c) trend (cause: equipment influences, aging of reagents, etc.) [141];
d) many entries close to the control limits (s enlarged).

2.6.7.1.5 Applications of the Shewhart Chart

The Shewhart chart is applied to analytical quality assurance in three different forms:

- mean value control charts,
- blank value control charts,
- recovery rate control charts.

2.6.7.1.5.1 Mean Value Control Charts

The mean value control chart corresponds to the original form of the Shewhart chart; however, in contrast to industrial product quality control, it is mostly applied to single values in analytical chemistry.

A mean value control chart serves mainly to validate the precision of an analytical process. Since systematic changes such as trends can also be detected, the accuracy may also be monitored to a limited extent.

Calibration parameters such as slope and intercept, in so far as they are determined daily, can also be tested by means of the mean value control chart. To this end, it is possible to use one abscissa (*x*-axis) for all parameters and to arrange the two ordinates above one another on one page [117].

The choice of control samples is dependent on the quality assurance situation in question and has no influence on the construction of a control chart (see Section 2.6.7.1.3).

Control samples for a mean value control chart
The following may be used as control samples for a mean value control chart (see also Section 2.5.2):

- **Pure standard solution**
  Advantage: Easy to prepare in any desired concentration.
  Disadvantages: Not representative of routine samples. Matrix effects cannot be detected.
  If an original standard sample from the calibration is used, systematic errors which appeared previously in the calibration cannot be detected either (lack of independence).

- **Synthetic samples**
  Advantage: Can be produced by the analyst as needed.
  Disadvantages: Preparation is very complicated for more complex matrices since the sample matrix must be simulated. The sample must be stable over time.

- **Reference samples**
  (stable natural samples that have been precisely analyzed using other analytical procedures; they may derive, for example, from interlaboratory, or round robin, tests.)
  Advantage: More economical than a certified reference material, but also independent.
  Disadvantage: Matrix and analyte content may not be representative.

- **Certified reference material**
  Advantage: An independent sample with a known expected value.
  Disadvantages: Relatively expensive and covers only a certain sample matrix, if at all. The analyte content is also not necessarily representative. May be used, however, for monitoring measurement instruments.

2.6.7.1.5.2 **Blank Value Control Charts**
The blank value control chart is a special application of the mean value control chart. It provides special information on the reagent and measurement systems used [105]. A particular feature is that measured values and not results (e.g., concentrations) are entered into the blank value control chart.

a) Scope of the Blank Value Control
Before establishing the decision criteria, a statement must be made concerning the scope of the blank value analyses to be performed.

In order to be correct, the simultaneous determination of the blank value would be required for each analysis; this requirement is however rarely fulfilled because it is very labor- and cost-intensive. It appears sensible to determine a minimum of two blank values ($n = 2$) per analysis series, whereby one is determined at the beginning and the other at the end of the series. Such an arrangement has, in addition to the precise determination of the blank value, the advantage that a drift check (see Section 1.7.1) can also be performed.

The arithmetic mean of n individual blank values per analysis series can be used to maintain the blank value control chart. The limits are determined using the standard deviation of the mean:

$$s_{\bar{x}} = \left(\frac{s}{\sqrt{n}}\right)$$

b) Maintaining a Blank Value Control Chart
The control and warning limits are determined as in the case of the $\bar{x}$-chart, i.e., from the results of analyses carried out in the preliminary period.

All further decision criteria, i.e., proof of high or low mean blank values and detection of trends, are the same as those for mean value control charts (see Section 2.6.7.1.5.1).

2.6.7.1.5.3 Recovery Rate Control Charts

a) General
Analyses of control samples and the management of mean value control charts give no indication of errors caused by matrix influences in the sample [17]. For analyses in which particular emphasis is placed on the specificity of the process (and therefore on the accuracy of the results), errors caused by interference from other substances contained in the sample matrix need to be as low as possible.

The analytical process may be tested for matrix influences by determining the recovery rate (RR) of spiked actual samples.

However, the recovery rate detects only matrix-dependent proportional systematic errors; constant systematic errors remain undetected when spiking (see Section 1.4.2.2). Therefore, the determination of the recovery rate is limited as a control of accuracy.

In exceptional cases, trueness may be checked by calculating the recovery rate, but then only if synthetic samples or reference samples or certified reference materials having known expected values (x_{exp}) are used.

The analysis of such certified reference materials produces an analysis value x_{actual}. The recovery rate can then be calculated as follows:

$$RR = \left(\frac{x_{actual}}{x_{exp}}\right) \cdot 100\% \tag{66}$$

The quality target should be a recovery rate of approximately 100%, since only then are systematic matrix influences completely absent.

b) Establishing the Spiking Quantity and Concentration
The spiking quantity or spiking concentration decisively influences the informational value of the recovery analyses. Spiking with too little material will lead to

too small a difference between the results for the spiked and unspiked samples. Therefore, in the literature it is recommended that the spiking concentration should at least correspond to the analysis value of the unspiked sample [11], i.e., the spiking should lead to an approximate doubling of the sample concentration.

In spiking experiments, the analysis value of the spiked sample must not exceed the upper range limit of the analytical process, i.e., all work must be within the verified linear range.

The spiking substance must be suited to the sample, i.e.,
- similar solubility,
- similar reaction behavior,
- similar oxidation state, etc.

Furthermore, it is important that the spiking substance is added directly to the sample as a solid (where possible) or as a very concentrated solution.

This should prevent:
1. a significant change in the sample matrix (e.g., if too large a volume of water or other matrix is added),
2. additional errors, such as dilution errors.

If spiking changes the volume of the sample by more than 0.5%, which can easily happen at concentrations of mg/l and more, then this volume error must be accounted for when calculating the recovery rate, RR (see Section 2.5.2.4).

c) Control Samples for the Recovery Rate Chart

The control sample types listed in Section 2.6.7.1.5.1 may be used for the recovery rate chart. RR is the quotient of $\hat{x}$ over x_{exp}:

- pure standard solutions,
- synthetic samples,
- reference samples,
- certified reference materials, all not spiked.

While using actual samples for mean value control charts is only sensible in exceptional cases, the recovery rate chart, on the other hand, is based almost entirely on spiked actual samples [176]:

Spiked natural samples from the analysis series in question

    Advantage: The sample requires no additional acquisition costs. It is representative of the analysis series in question. There are no stability problems, since no control sample must be stored.

    Disadvantage: Only proportional systematic errors can be detected. Constant systematic errors, which can occur during analysis of the bulk (or gross) sample, are not quantifiable in this way.

With such a multitude of different matrices (drinking water, waste water, surface water, blood, urine, tissue samples, plant extracts, foodstuffs, etc.), it is absolutely necessary to maintain one's own recovery rate chart for each matrix type. To

2.6 The Control Chart System

do this, an actual sample of each matrix type should be chosen daily for the recovery experiment. The concentration of these samples should, if possible, lie in the lower half of the process range or the actual samples must be diluted accordingly.

d) Construction of a Recovery Rate Control Chart

The recovery rate control chart is equivalent in set-up and decision criteria to the mean value control chart.

The mean recovery rate, $\overline{RR}$, and its standard deviation, s_{RR}, are calculated after a sufficiently long preliminary period (e.g., $N = 20$).

$$\overline{RR} = \frac{1}{N} \cdot \sum_{i=1}^{N} RR_i \tag{67}$$

$$s_{RR} = \sqrt{\frac{1}{N-1} \sum_{i=1}^{N} \left(RR_i - \overline{RR}\right)^2} \tag{68}$$

The warning and control limits are the $2 \cdot s$ and $3 \cdot s$ ranges, respectively:

- upper control limit, $UCL = \overline{RR} + 3 \cdot s_{RR}$
- lower control limit, $LCL = \overline{RR} - 3 \cdot s_{RR}$
- upper warning limit, $UWL = \overline{RR} + 2 \cdot s_{RR}$
- lower warning limit, $LWL = \overline{RR} - 2 \cdot s_{RR}$

As for the mean value control chart, the central line of the chart represents the mean recovery rate. The control limits are then drawn at a distance of $\pm 3 \cdot s$ and the warning limits at a distance of $\pm 2 \cdot s$ from this line [89, 147]. As an accuracy check, a recovery rate of 100 % may be entered as the central line [17].

The out-of-control rules for the Shewhart chart are also used to evaluate the recovery rate control chart.

2.6.7.2 R-Chart (Range Control Chart)

2.6.7.2.1 General

While the Shewhart chart shows how well mean values of subgroups or single values correlate with the grand average (process mean), it does not provide any information about the distribution of individual results within and between the subgroups. In contrast, the R-chart (R = range) serves above all the purpose of precision control. For this, range is defined as the difference between the largest and smallest single results of multiple analyses.

An R-chart offers the following control possibilities:

1. Precision control
   A fundamental requirement for the use of Shewhart charts is the constancy of the standard deviations [214]. An R-chart allows the detection of significant differences between the standard deviations of the subgroups [3] and thus represents a test of the variance homogeneity.

2. Limited accuracy control – drift control
Trends within a series and equipment drifts may be monitored by means of an R-chart which has been modified to a difference chart (see Section 2.6.7.3) [109].

The use of R-charts is only meaningful if the control limits are not set by outside factors, but result from the imprecision of the analysis [214].

2.6.7.2.2 Construction of an R-Chart

In order to construct an R-chart, the following quantities must be known or calculated:

- number n of repeated measurements per subgroup (n_i), at least $n = 2$,
- number N of subgroups, series,
- R_i, range of subgroup i,
- $\bar{R}$, mean value of ranges,
- (s^2, variance of the entire measurement),
- upper warning limit, UWL,
- lower warning limit, LWL,
- upper control limit, UCL,
- lower control limit, LCL.

First, the ranges, R_i, of all subgroups are determined and combined as $\bar{R}$:

$$R_i = \text{largest value} - \text{smallest value of a subgroup } i \text{ consisting of } n \text{ single measurements} \quad (69)$$

$$\bar{R} = \frac{\sum R_i}{N} \quad (70)$$

$\bar{R}$ forms the "central line" of the R-chart (see Figure 2-12).

The respective warning and control limits are obtained from the mean range, $\bar{R}$, by multiplying it by a factor D, which is a function of the number of multiple determinations, n, and the significance level:

$$\begin{aligned}
\text{LCL} &= D_{\text{Clo}} \cdot \bar{R} \quad (D \text{ factors: see Table 2-5})\\
\text{UCL} &= D_{\text{Cup}} \cdot \bar{R}\\
\text{LWL} &= D_{\text{Wlo}} \cdot \bar{R}\\
\text{UWL} &= D_{\text{Wup}} \cdot \bar{R}
\end{aligned}$$

One may choose between two different significance level combinations:

	Warning limit	Control limit	References
Combination a	95%	99%	[23]
Combination b	95%	99,7%	[3, 4, 14, 109, 141]

Combination 'b' is based on the same $\pm 2 \cdot s$ and $\pm 3 \cdot s$ significance limits as in the case of the Shewhart chart and is therefore generally preferred.

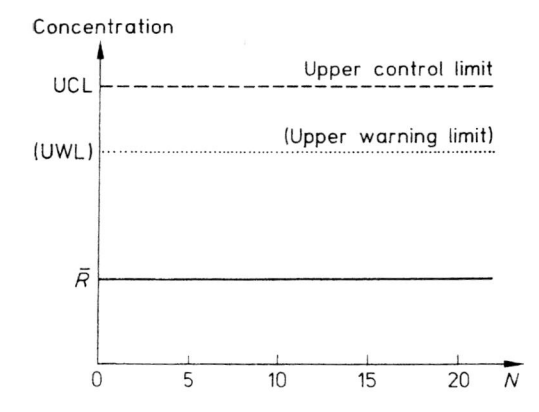

Fig. 2-12 R-chart (range chart) with lower limits at zero.

Table 2-5 D-factors for the calculation of R-chart limits.

n	P = 95% [23]		P = 99% [23]		P = 99.7% [3, 4]	
	D_{Wlo}	D_{Wup}	D_{Clo}	D_{Cup}	D_{Clo}	D_{Cup}
2	0.039	2.809	0.008	3.518	0.000	3.267
3	0.179	2.176	0.080	2.614	0.000	2.575
4	0.289	1.935	0.166	2.280	0.000	2.282
5	0.365	1.804	0.239	2.100	0.000	2.115
6	0.421	1.721	0.296	1.986	0.000	2.004
7	0.462	1.662	0.341	1.906	0.076	1.924
8	0.495	0.617	0.378	1.846	0.136	1.864
9	0.522	1.583	0.408	1.798	0.184	1.816
10	0.544	1.555	0.434	1.760	0.223	1.777
11	0.562	1.531	0.456	1.729	0.256	1.744
12	0.578	1.511	0.475	1.702	0.284	1.716
13	0.592	1.494	0.491	1.679	0.308	1.692
14	0.604	1.480	0.506	1.659	0.329	1.671
15	0.615	1.467	0.519	1.642	0.348	1.652
16	0.625	1.455	0.531	1.626	0.364	1.636
17	0.634	1.445	0.542	1.612	0.379	1.621
18	0.642	1.435	0.552	1.599	0.392	1.608
19	0.649	1.427	0.561	1.587	0.404	1.596
20	0.656	1.419	0.569	1.577	0.414	1.586

The lower limits may be omitted for relatively stable processes, i.e. relatively reliable analysis processes. They should, however, be retained for equipment monitoring [139].

2.6.7.2.3 Out-of-Control Situations

The decision criteria described for Shewhart charts can only be conditionally applied for R-charts.

For example, it is not reasonable to assume an out-of-control situation if seven consecutive values lie underneath the center line. This could possibly imply just the opposite – an improvement of the precision. If such a situation were to continue, then it is recommended that the control limits be recalculated.

2.6.7.2.4 Decision Criteria for R-Charts
An out-of-control situation exists if

- one R_i value lies above the upper control limit ('1' in Figure 2-13),
- one R_i value lies below the lower control limit (valid only if LCL > 0),
- seven consecutive values show an ascending ('2' in Figure 2-13) or descending trend ('3' in Figure 2-13),
- seven consecutive values lie above the range mean, $\bar{R}$ ('4' in Figure 2-13).

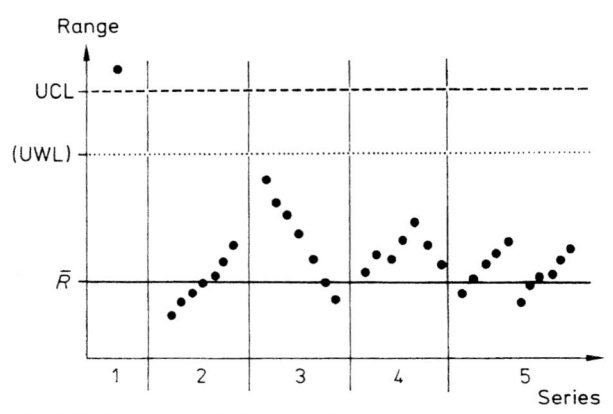

Fig. 2-13 Out-of-control situations for R-charts.

So, if just one R_i in the preliminary period lies outside of the upper control limit, all quantities required for the construction of an R-chart must be recalculated.

Cyclical movements [ascending ('5' in Figure 2-13) or descending] of the ranges indicate influences resulting from the maintenance schedule of the instruments or aging of the reagents [141]. However, this is not an out-of-control situation.

2.6.7.2.5 Applications of R-Charts
For practical applications in analytical laboratories, the R-chart appears only in its simplest form, i.e., only one duplicate determination per analysis series. A larger number of parallel determinations would indeed result in a more sensitive precision control, however the effort required for $n > 2$ is usually not acceptable.

2.6.7.2.6 Control Samples for R-Charts
Constructing R-charts by means of standard samples or control samples with known content is of little value since the mean value control chart can provide sufficient information instead. Analysis of standard reference materials also has definite disadvantages that are not seen with actual samples. Matrix influences, devia-

tions in the stability of the sample, etc., are only registered in quality assurance through the use of actual samples. The precision is therefore verified under conditions close to reality.

Whitby [from 179] logically recommends that duplicate determinations be performed on actual samples as often as possible. At the beginning of each analysis series, one of the actual samples is declared a "control sample" and it is analyzed again at the end of the series (in all, a duplicate determination).

It must be taken into account that the actual samples lie within a concentration range, since the standard deviation and also the range are often concentration-dependent (see Table 2-6) [89, 179]. Should an R-chart be maintained independently of the concentration, then the percentual relative range must be related to it [89]. At the least, changing the measuring or display range will result in a significant change of precision.

Example of Concentration-Dependent Ranges

Table 2.6 Concentration-dependent ranges.

x_{i1}	x_{i2}	R_i	$\bar{x}_i$	$\dfrac{R_i}{\bar{x}_i} \cdot 100\%$
11.2	11.0	0.2	11.1	1.80%
23.5	24.0	0.5	23.75	2.11%
114.3	112.0	2.3	113.15	1.24%
69.3	70.7	1.4	70.0	2.00%
78.7	80.3	1.6	79.5	2.01%
100.5	102.5	2.0	101.5	1.97%
203.4	198.8	4.6	201.1	2.29%
54.7	55.8	1.1	55.25	1.99%
150.1	153.1	3.0	151.6	1.98%
90.8	89.0	1.8	89.9	2.00%

The R values listed in Table 2-6 have been represented graphically using the means (see Figure 2-14).

The R_i values are obviously dependent on the concentration. The correlation coefficient amounts to

$$r = \frac{\sum (\bar{x}_i - \bar{\bar{x}}) \cdot (R_i - \bar{R})}{\sqrt{\sum (\bar{x}_i - \bar{\bar{x}})^2 \cdot (R_i - \bar{R})^2}} = 0.9941 \tag{71}$$

A dependence between the range (R-values) and the concentration ($\bar{x}$-values) can be proven statistically if the correlation coefficient r is significantly different from zero. To test this, the value of r at $f = n - 2$ degrees of freedom is compared with the threshold value $r(P = 99\%, f)$ (see Table 2-7). If $|r| > r(P = 99\%, f)$, then a dependence exists.

Range

[Figure: scatter plot with linear fit showing range vs. mean of duplicate determination, x-axis 0 to 240, y-axis 0 to 5]

Mean of the duplicate determination

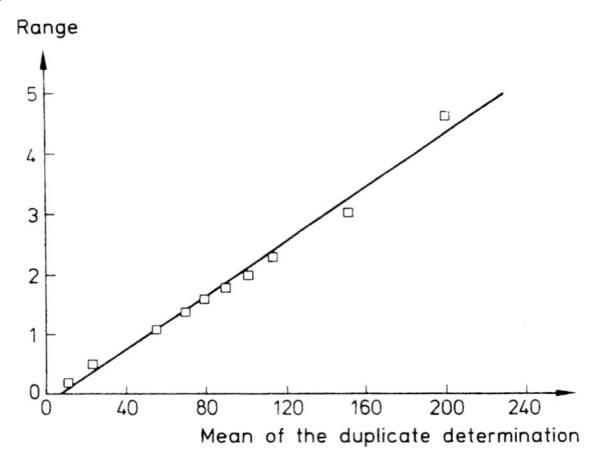

Fig. 2-14 Dependence of the ranges on concentration.

Table 2-7 Threshold values $r(P = 99\%, f)$ for testing the correlation coefficient [78].

f	$r(P = 99\%, f)$	f	$r(P = 99\%, f)$
1	1.00	16	0.59
2	0.99	17	0.58
3	0.96	18	0.56
4	0.92	19	0.55
5	0.87	20	0.54
6	0.83	25	0.49
7	0.80	30	0.45
8	0.77	35	0.42
9	0.74	40	0.39
10	0.71	45	0.37
11	0.68	50	0.35
12	0.66	60	0.33
13	0.64	70	0.30
14	0.62	80	0.28
15	0.61	100	0.25

In the example, $r = 0.9941$ is larger than $r(P = 99\%, f = 8) = 0.77$, which certifies that there is a correlation between range and concentration.

Note: A separate *R*-chart must be maintained for each type of matrix (e.g., waste water, surface water), since different matrices can have very different influences on the precision of the results.

A difference chart (see Section 2.6.7.3) is recommended to check drift of actual samples [179].

2.6.7.2.7 $\bar{x}$-R Combination Charts

The $\bar{x}$-R combination chart is probably the most useful control chart used in industrial quality assurance [3]. It consists of an $\bar{x}$-chart and an R-chart together, arranged so that the mean value and range for *one* given subgroup are positioned above one another on the graph (see Figure 2-15).

This process allows different changes to be recorded simultaneously on one chart. The $\bar{x}$-chart reacts sensitively to changes in the mean values of the subgroups; in contrast, the R-chart provides information about too large a distribution within a subgroup.

The primary advantage of this method is that it enables one to decide whether the deviation between subgroups is significantly larger than the deviation within a subgroup. In this situation, the R-chart is in control and the $\bar{x}$-chart indicates "out-of-control" situations [179]. This occurs frequently with certain chemical processes, indicating insufficiently controlled variables (e.g., temperature).

If the opposite situation occurs, i.e. the $\bar{x}$-chart is in control and the R-chart is out of control, or if a trend is spotted, this indicates a change in the individual variances [141].

If the mean values tend to always move in the same direction as the range, it could be a "skewed" distribution [141] (the same is true for continuous movement in the opposite direction).

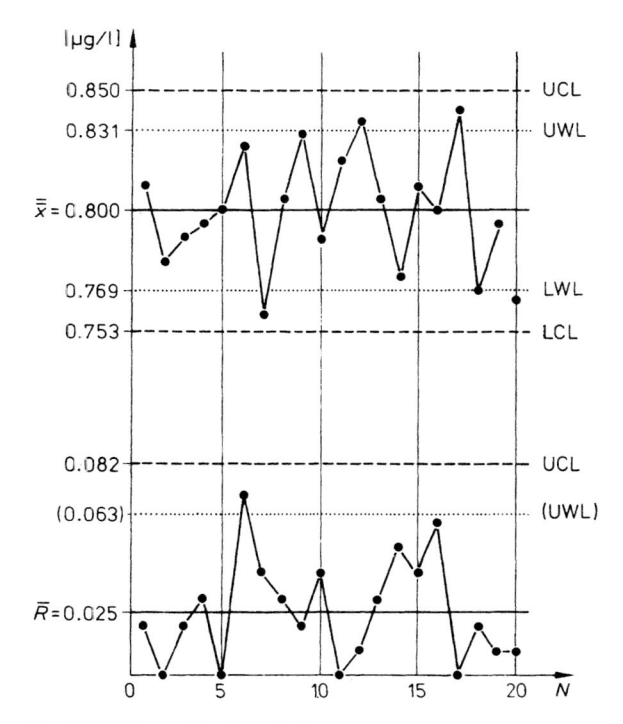

Fig. 2-15 $\bar{x}$-R combination chart.

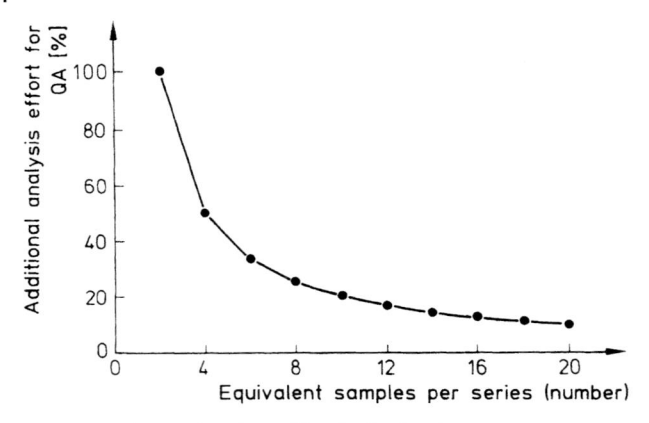

Fig. 2-16 Additional analysis effort for the $\bar{x}$-R chart.

The $\bar{x}$-R chart only makes sense if the same control sample is used for both range and mean value control. A control sample (synthetic or natural) that remains stable over a long period of time is required.

The maintaining of a matrix-representative $\bar{x}$-R chart is especially recommended for the regular analysis of multiple similar samples within an analysis series. The minimum required duplicate determination of a control sample per matrix increases the total analysis effort significantly and is not justifiable if only a few of the samples within the series have the same matrix (see Figure 2-16). For example, the additional control analysis effort required decreases to $\leq 10\%$ only for a series of 20 or more samples.

2.6.7.3 Difference Chart [17, 179]

While the R-chart uses the absolute value of the difference between the largest and smallest analysis values, the difference chart uses both the size of the difference and its sign. The objective is specifically the detection of systematic errors that first appear *during* an analysis series. For a difference chart, an actual sample from each series is analyzed twice. As many analyses as possible from the series should lie between the two determinations.

In each case, the reference value is the first value measured. The differences $d_i = x_{i2} - x_{i1}$, whereby the first value must always be subtracted from the second value, are entered on a control chart while observing the appropriate signs. The central line of this chart represents the expected value of the differences, i.e., the value zero. Under optimal conditions, the differences must be distributed around the zero-line according to their signs. If a drift is present (for example, from the equipment), x_2 will become ever larger (or smaller) than x_1, which is documented by a lop-sided position of the differences above (or below) the zero line (see Figure 2-17).

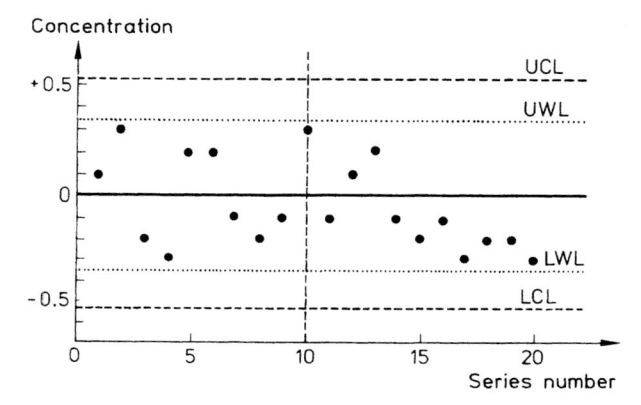

Fig. 2-17 Difference chart showing the presence of drift.

2.6.7.3.1 Calculation of Control Limits

The control limits are calculated:

a) From the standard deviation, s_d [17]:

$$s_d = \sqrt{\frac{\sum (d_i - \bar{d})^2}{N - 1}}$$

central line: should be 0
warning limit: $\pm 2\, s_d$
control limit: $\pm 3\, s_d$

Since s_d is dependent on the concentration, this control chart may only be used for samples which lie within a limited concentration range.

b) Using the mean range [179]:

$$\bar{R} = \frac{\sum R_i}{N}$$

mean: $= 0$
warning limit: $1.77 \cdot \bar{R}$ (comparable to $2 \cdot s$)
control limit: $2.65 \cdot \bar{R}$ (comparable to $3 \cdot s$)

2.6.7.3.2 Control Samples for the Difference Chart

An absolute requirement for maintaining a difference chart is the use of stable control samples. Their choice is determined by the monitoring objective.

If only the measuring instrument (detection step) is to be monitored, then unknown matrix errors must be ruled out. Only standards which undergo the actual measurement may be considered as control samples. This procedure is comparable to measurements for instrument adjustment.

Stable actual samples are required to monitor the entire analytical process. As an alternative to synthetic samples and stable natural samples, which serve as control samples over a longer period of time, an actual sample from the analysis series in question may be used (similarly to range control). However, it must be guaranteed that the sample undergoes no physical, chemical or biological changes during the time period between the two control analyses.

2.6.7.4 Standard Deviation Chart (s-Chart)

This chart [23] is used in a similar manner to the R-chart to monitor the distribution of the basic entity.

The s-chart, however, is better suited for subgroup sizes of $n > 10$ than is the R-chart. Used in combination with a mean value chart, the s-chart is able to detect systematic errors quickly [4].

2.6.7.4.1 Calculation of the Central Line and Control Limits for the s-Chart

The *central line* is entered as the mean standard deviation (s_w) [113]:

$$s_w = \sqrt{\frac{\sum_{i=1}^{N} f_i \cdot s_i^2}{\sum_{i=1}^{N} f_i}} = \sqrt{\frac{\sum_{i=1}^{N} (n_i - 1) \cdot s_i^2}{\sum_{i=1}^{N} (n_i - 1)}} \tag{72}$$

or for duplicate determinations ($n_i = 2$):

$$s_w = \sqrt{\frac{\sum s_i^2}{N}} \tag{73}$$

The control limits are obtained from the χ^2 distribution using the table values $\chi^2\left(n-1; \frac{\alpha}{2}\right)$ and $\chi^2\left(n-1; 1-\frac{\alpha}{2}\right)$ (see Table 2-8):

$$UCL = s_w \cdot \sqrt{\frac{1}{n-1} \cdot \chi^2\left(n-1; \frac{\alpha}{2}\right)} \tag{74}$$

$$LCL = s_w \cdot \sqrt{\frac{1}{n-1} \cdot \chi^2\left(n-1; 1-\frac{\alpha}{2}\right)} \tag{75}$$

Table 2-8 Threshold values of the χ^2 distribution for $\alpha = 1\%$.

$n-1$	$1 - \dfrac{\alpha}{2} = 99.5\%$	$\dfrac{\alpha}{2} = 0.5\%$
1[a]	$3.93 \cdot 10^{-5}$	7.879
2	$1.00 \cdot 10^{-2}$	10.60
3	$7.17 \cdot 10^{-2}$	12.84
4	0.207	14.86
5	0.412	16.75

[a] corresponds to a duplicate determination ($n = 2$).

2.6.7.5 Target Value Charts [136] [1]

Control charts usually proceed on the basis of a normal distribution of values and so their characteristics are statistically determined.

Another possibility is to run control charts on the basis of empirically derived or statutorily required exclusion limits, i.e. as so-called target value charts. Mean value, blank value, recovery rate, and range control charts can all be used as target value charts.

A target value chart is particularly appropriate if:
- no normal distribution of the control values is present (e.g., blank values),
- no sufficient database is available for the determination of statistical characteristics (e.g., rarely conducted analyses), or
- a particular quality of the analytical results should be guaranteed in order to meet the specifications of an exclusion limit [88, 183].

2.6.7.5.1 Control Samples for Target Value Charts

To establish a target value chart, it is possible to use the same control samples as in the case of the Shewhart chart (see Section 2.6.7.1) or the range chart (see Section 2.6.7.2).

2.6.7.5.2 Determining the Exclusion Limits

With target value charts, there are no warning or control limits, only an upper and a lower exclusion limit. These exclusion limits can be derived from:
- legal specifications,
- parameter-related standards and AQC instruction cards,
- the (lowest) guaranteed laboratory-internal precision and trueness for a measurement,
- the assessment of laboratory-internal data series.

---

[1] For example, in Germany the Länder Arbeitsgemeinschaft Wasser (LAWA) or the German Working Group of the Federal States on water issues recommends the use of target value charts for water analysis: LAWA AQS leaflet A-2 [113].

The target value chart is constructed with the aid of the exclusion limits, and so a preliminary period for the determination of control chart limits may be dropped.

For range-target value charts, only an upper exclusion limit needs to be established.

2.6.7.5.3 Out-of-Control Situations
The analysis procedure is out of control when the stipulated exclusion limits are transgressed.

Target value charts are nevertheless suitable for detecting trends even if no "out-of-control" situation has occurred.

2.6.7.6 Cusum Chart

2.6.7.6.1 Principle of the Cusum Chart
The cumulative sum, S (cusum), is understood as the sum of the deviations from a target value [202]. The target value may, for example, correspond to the mean value of a control sample determined in the preliminary period. This mean value, also known as the reference value, k, is subtracted from every control analysis result, x_i, and the difference is added to the sum of all previous differences [139].

Cumulative sums:

$$S_1 = (x_1 - k)$$
$$S_2 = S_1 + (x_2 - k)$$
$$S_3 = S_2 + (x_3 - k) \tag{76}$$
$$\cdot$$
$$\cdot$$
$$S_N = S_{N-1} + (x_n - k) = (\sum x_i) - nk$$

The cusum value S_N (ordinate) is then plotted against the number of observed results N (abscissa) on a control chart.

To determine an out-of-control situation, the slope of the cusum line is evaluated (see Figure 2-18). If the process is in control, the results of the control analyses vary randomly around the reference value k, and the cumulative sums are distributed around the value 0. A change in the process leads to an increase or decrease in the cumulative sum, the slope of the cusum line changes, and an out-of-control situation is detected.

2.6.7.6.2 Purpose and Applications of the Cusum Chart
The cusum chart was introduced into industrial quality assurance by E. S. Page [175] in 1954. This chart is a further development of the Shewhart chart, whereby single results are no longer entered but instead the summation of the deviations of the single results from a set value.

Therefore, each new entry contains not only information about the current status but also about past analysis values of the analytical process in question [81].

In this manner, changes that would not lead to an out-of-control situation on an $\bar{x}$-chart can be more easily detected. This is especially advantageous for processes with a large variance [211].

The cusum technique is especially suited to the following applications:

- recognition of a systematic change or shift in the mean value of a process in progress [210],
- determination of the order of magnitude by which the mean value has changed [210],
- determination of the point in time at which the change occurred [210],
- short-term prediction of the future mean value [216].

2.6.7.6.3 Establishing a Cusum Chart
A cusum chart should be dimensioned such that it reacts sensitively to the smallest deviation in the mean value that is seen as important in practice. This reaction should be reflected in a clearly visible change in the slope of the cusum line. In the case of an "in-control process", the slope of the line is decisively dependent upon the chosen reference value k; in the case of an "out-of-control process", the clearness of the change in the slope is dependent on the scale of the y-axis (cusum axis). Therefore, one must keep in mind not only a correct choice of the reference value k, but also of the scale factor, w.

2.6.7.6.4 Choice of the Reference Value k
If a certified reference material is available for use in quality assurance, then the given standard concentration is equivalent to the reference value k. If this is not the case, the concentration of the control sample must be determined in a preliminary period (as for an $\bar{x}$-chart, *vide supra*). The estimate of the mean value should be based on at least 20 analyses.

If the reference value k is imprecisely determined, then the cusum values do not fluctuate about the value 0, i.e., the cusum line rises or falls continuously. An example of this is shown in Figure 2-18.

If the "correct" reference value is not chosen, on the one hand the upper or lower limits of the cusum chart may be reached very quickly, but on the other hand small changes in slope are more difficult to recognize.

2.6.7.6.5 Scale of the y-Axis (Cusum Axis)
The abscissa and ordinate are first assigned equidistant divisions (each designated as 1 unit).

The scale of the axes is determined by the scale factor, w. This indicates which cusum value corresponds to a unit on the cusum axis. Normally, w is expressed as "q number of standard deviations". For this, the standard deviation s is determined from the preliminary period standard deviation s' (see also $\bar{x}$-chart; see Section 2.6.7.1.5.1).

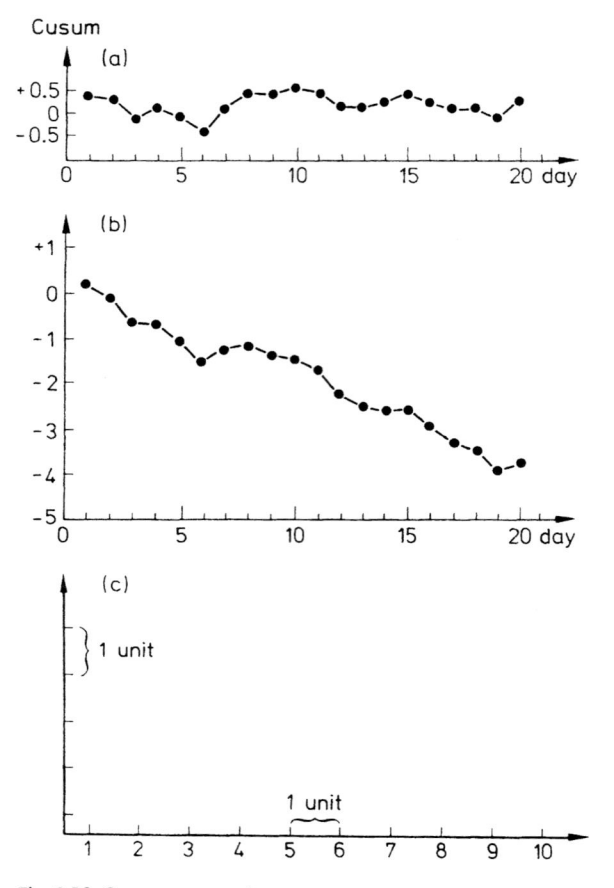

Fig. 2-18 Cusum progression:
a) cusum progression for a properly chosen reference value $k = 6.0$,
b) cusum progression for an incorrect reference value $k = 6.2$,
c) scale of the cusum axis.

For only one control analysis per series, $s = s'$; if the cusum chart is constructed with mean values from each of n analyses, then the standard deviation of the mean values is included in the scale:

$$s = \frac{s'}{\sqrt{n}}$$

Both the scale of the cusum axis and the correct (or skillful) choice of the reference value contribute to making a cusum chart that is easy to handle. The cusum axis dimensions should neither be too far apart (see Figure 2-19b) nor too close together as a change in the slope is difficult to recognize in both cases.

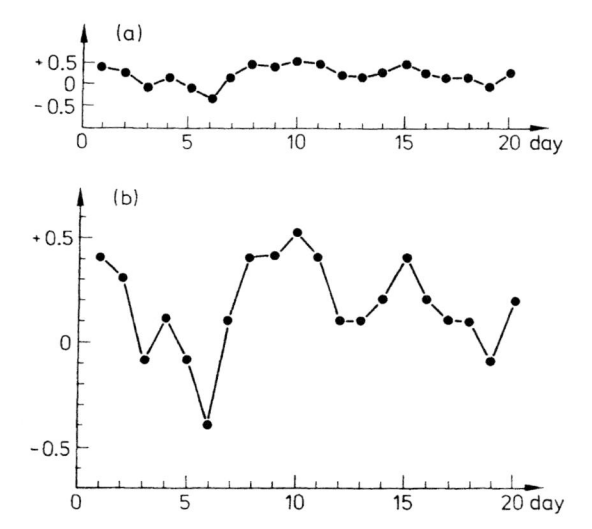

Fig. 2-19 Influence of the cusum axis scale;
a) suitable choice of scale ($w = 2s$),
b) scale increments too far apart.

Goldsmith and Whitfield [106] and Ewan [96] recommend, for example, a scale factor of $w = 2 \cdot s$ ($q = 2$) to $w = 1 \cdot s$ ($q = 1$) for use with graphical decision criteria. If the distance between two entries on the x-axis (e.g., one day) is defined as a unit, then the same distance on the cusum axis is represented by $2 \cdot s$. The cusum line rises at $45°$ if two consecutive values differ by $2 \cdot s$.

2.6.7.6.6 Determination of an Out-of-Control Situation
In order to determine an out-of-control situation, there are three decision criteria for cusum charts:

1. Visual decision criterion.
2. The V-mask as decision criterion.
3. Numerical decision criterion (this corresponds to the V-mask under standardized conditions).

2.6.7.6.6.1 Visual Decision Criterion
A deviation of the actual from the required mean value leads to a change in slope. This change can be easily determined visually if the chart is properly dimensioned and the process is reliable. However, only slight changes in the cusum progression make a visual interpretation difficult [109, 211].

In order to make sound decisions, which can be reproduced by others, it is advantageous to use an objective decision criterion.

2.6.7.6.6.2 V-Mask

The V-mask, developed by Barnard [6] in 1959, is a two-sided statistical test that can determine positive and negative deviations from a mean value. Since it is possible to consider the previous cusum progression, the V-mask technique combines visual interpretation with objective test criteria.

The V-mask is defined by the two parameters d and θ (see Figure 2-20).

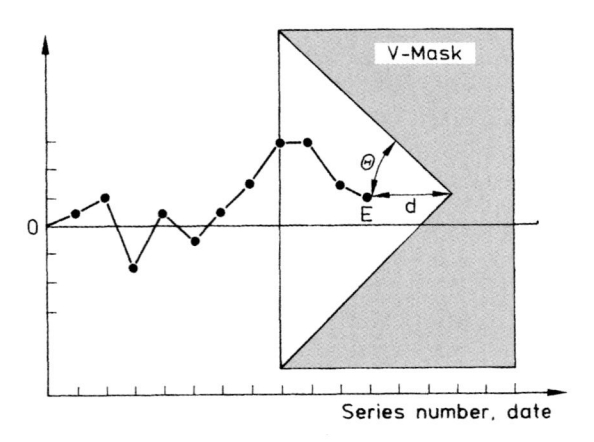

Fig. 2-20 V-mask and parameters d and θ.

The leading distance d represents the distance from the vertex of the V-mask to the most recent entry on the chart; d is expressed in abscissa units (e.g., days). θ is the angle between the arms of the mask and the horizontal drawn through the mask vertex. After establishing these two parameters, the V-mask may be drawn on transparent film or cut out of cardboard.

The V-mask is then positioned on the cusum chart at a distance d from the most recent entry, so that the vertex (placed horizontally) points forward. For each new entry, the mask is shifted so that the point E comes to lie on the new cusum value. This represents a shifting of one abscissa unit. It should be noted that the V-mask may *not* turn its vertex around, i.e. the leading distance d remains parallel to the x-axis.

An out-of-control situation is indicated if the cusum line crosses one of the arms of the V-mask. The larger the values of θ and d, the more infrequently an out-of-control situation arises.

If the cusum line cuts through the upper arm of the mask, then the mean value has decreased. The mean value has increased if the line crosses the lower arm (see Figure 2-21).

The first cusum value that lies outside of the V-mask indicates the point in time at which the out-of-control situation appeared. This information can be very helpful when searching for the cause of the error. If the error is discovered and corrected, then the cumulative summation begins again at zero.

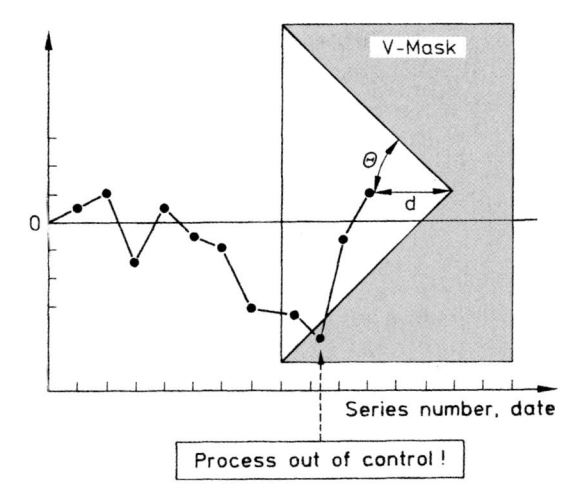

Fig. 2-21 Out-of-control situation.

2.6.7.6.6.3 Determination of the V-Mask Parameters θ and d

There are several possible ways to determine the V-mask parameters θ and d, all of which are based on the mathematical-statistical procedure for the determination of sequential test plans [139, 141, 216]. Either the simplified approximation calculation of V-mask parameters of Juran ([141], see below) or, using the ARL, a selection from a collection of previously calculated V-masks both lend themselves to practical application.

The quality assurance objective forms the basis for the V-mask parameters: which systematic deviation D (expressed in standard deviation units) of the process mean from the reference value should be recognized

- with what probability of error,
- in what time frame,
- with what certainty from undetected error (β-error).

(1) Preferred methods: Ratzlaff [178] and Juran [141] give approximation equations for the parameters d and θ:

$$\theta = \arctan\left(\frac{D}{2 \cdot w}\right) \tag{77}$$

$$d = \frac{2 \cdot s^2 \cdot \ln \alpha}{D^2} \tag{78}$$

D is the smallest deviation that can be proven with a set degree of certainty;
$D = |x_{\max} - k|$ or $|D = x_{\min} - k|$
(D should be expressed as a multiple of the standard deviation s)
s standard deviation of the reference value, e.g., obtained from the preliminary period

w scale factor (e. g., $w = 2 \cdot s$)
α probability of an error of the first kind
(false alarm: an out-of-control situation is signalled even though the process is in control)
($\alpha = 0.0027$ corresponds to the $3s$ control limit of the $\bar{x}$-chart)

Ratzlaff [178] assumes a negligible error of the second kind (undetected alarm) ($\beta < 0.01$).

(2) Jardine [139] and other authors [96, 106, 210, 216] recommend choosing parameters using the average run length (ARL; see Section 2.6.3).

First, the ARL L_1 should be chosen for a deviation of the current mean value from the target value that is no longer tolerable, e. g., $L_1 = 10$.

The ARL L_0 should in any case be very high, e. g., $L_0 = 500$, if the process is in control, i. e., the probability of a false alarm must be very low.

Woodward and Goldsmith [106, 216] determined ARL curves for various V-mask parameter combinations. The four figures (a–d) in Figure 2-22 are taken from their monograph "Cumulative Sum Techniques".

Applied example:

$L_0 = 500$
$L_1 = 15$, for a $1 \cdot s$ deviation from the target value
Curve VII in Figure 2-22 b fulfills the requirements most fully.

Therefore, one obtains:

$d = 2$
$\tan \theta = 0.55$ (i. e., $\theta = 28.8°$)

Since the V-mask is a graphical method, the diagram scale must be taken into account. While the distance d remains uninfluenced by the scale, $\tan \theta$ must be fitted using the appropriate scale factor:

$$q = \frac{w_{actual}}{s}$$

$\tan \theta_{actual} = q \cdot \tan \theta_{table}$

Based on analytical experience, Haeckel [109] and Doerffel [77] recommend the following V-mask parameters:

Scale:
$w = 2 \cdot s$
$d = 8$ or $d = 10$
$\theta = 15°$ $\theta = 14°$

(3) Barnard [6] suggests testing different V-masks on old cusum charts and using the most suitable value in the future. However, the masks at one's disposal should be statistically certified, i. e., defined on the basis of the ARL (see above).

2.6 The Control Chart System

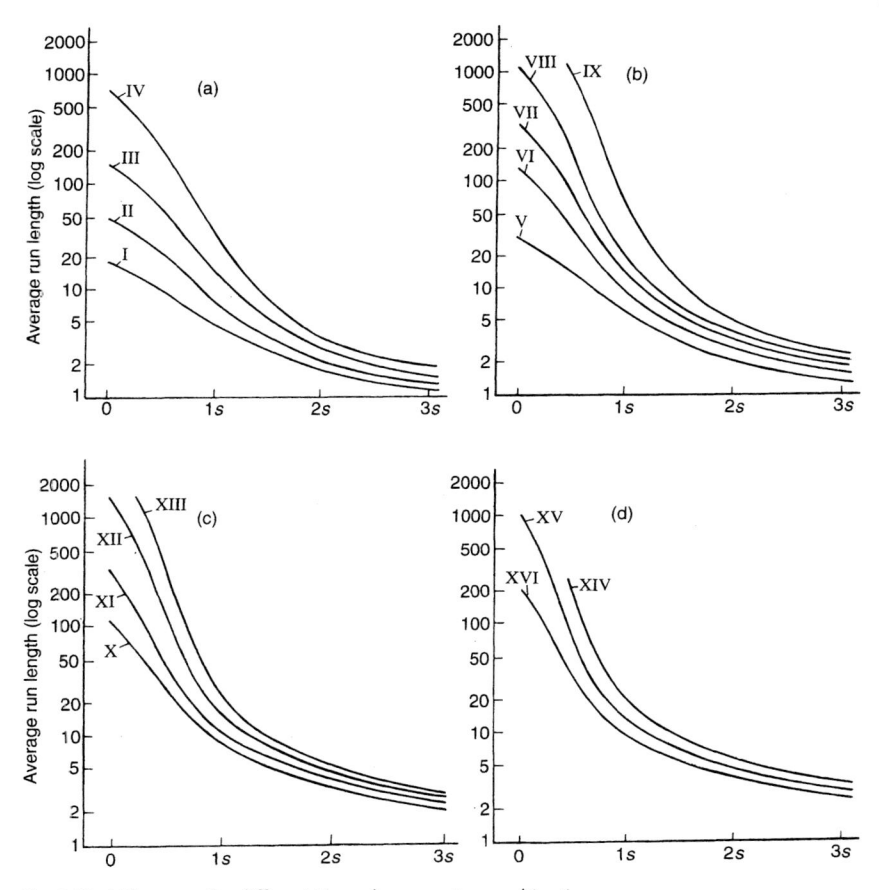

Fig. 2-22 ARL curves for different V-mask parameter combinations.

a) Curve	d	tan θ
I	1	0.500
II	1	0.600
III	1	0.700
IV	1	0.800

b) Curve	d	tan θ
V	2	0.400
VI	2	0.500
VII	2	0.600
VIII	2	0.700
IX	2	0.800

c) Curve	d	tan θ
X	5	0.300
XI	5	0.350
XII	5	0.400
XIII	5	0.450

d) Curve	d	tan θ
XIV	8	0.250
XV	8	0.300
XVI	8	0.350

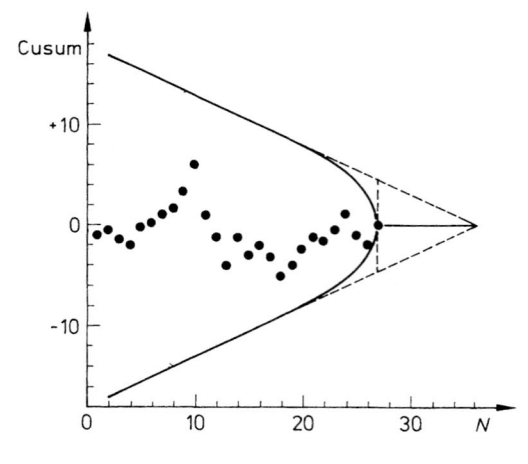

Fig. 2-23 Parabolic V-mask according to Lucas [148].

(4) Lucas [148] suggests the use of a parabolic mask (see Figure 2-23), which is only mentioned here to complete the listing

2.6.7.6.6.4 Numerical Decision Criteria

An alternative to the V-mask is the numerical decision criterion developed by Ewan and Kemp [95]. This technique [139, 202, 210, 216] requires no graphical representation of the cusum values. The calculation effort is small, as long as the results of the analytical process are satisfactory.

For $\bar{x}$-charts, exceeding the $3 \cdot s$ control limit signals an out-of-control situation. For the cusum technique, this limit represents the *decision interval*, i.e., the threshold values h act as action limits.

For the numerical technique, the cumulative sum is not created as long as the results are below a *decision limit K*. (The decision limit K should not be confused with the reference value k.)

To determine the parameters h and K, the following criteria must be established:

- D represents the smallest deviation of the mean from the target value that can no longer be tolerated.
- L_1 is the number of control experiments (ARL) after which the deviation should, at the latest, be detected (ARL; see Section 2.6.3).
- If positive and negative systematic deviations are detected, a two-sided test must be used, which in this case consists of a combination of two one-sided tests (deviation upward and deviation downward) [210].
- Additionally, the probability of the appearance of an error of the first kind (α-error, false alarm) should be very small.

In contrast to the V-mask, one- and two-sided questions are possible. The numerical decision criteria should be valid for values too high *and* too low (two-sided question). The decision limits K are calculated as follows from the grand average $\bar{\bar{x}}$

determined in the preliminary period and the smallest deviation D thereof that may no longer be tolerated:

$$K_{\text{upper}} = \bar{\bar{x}} + \frac{1}{2}D \tag{79}$$

$$K_{\text{lower}} = \bar{\bar{x}} - \frac{1}{2}D \tag{80}$$

The cumulative summation only begins when

$$\bar{x}_i > K_{\text{upper}}$$

or

$$\bar{x}_i < K_{\text{lower}}$$

If a negative systematic deviation is detected, then the value K_{lower} is subtracted from the respective subgroup mean $\bar{x}_i$:

$$S^- = \sum (\bar{x}_i - K_{\text{lower}}) \tag{81}$$

For a positive systematic error, S^+ is created:

$$S^+ = \sum (\bar{x}_i - K_{\text{upper}}) \tag{82}$$

The summation is continued until:

a) the decision limit K_{upper} or K_{lower} is obtained, i.e. the process goes out of control
or
b) $S^+ < 0$ or $S^- > 0$

The decision interval h can be determined from a nomogram [202, 210, 216] or by means of a table [202].

Application:

1. Determination of K according to Eqs. (79) and (80)
2. Formation of the quotient $\dfrac{|\bar{\bar{x}} - K| \cdot \sqrt{n}}{s}$
3. Choice of L_1
4. Linking of L_1 with the quotient from 2. in the nomogram
5. Reading of $\dfrac{h \cdot \sqrt{n}}{s} = ...$
6. Calculation of $h = \dfrac{... \cdot s}{\sqrt{n}}$

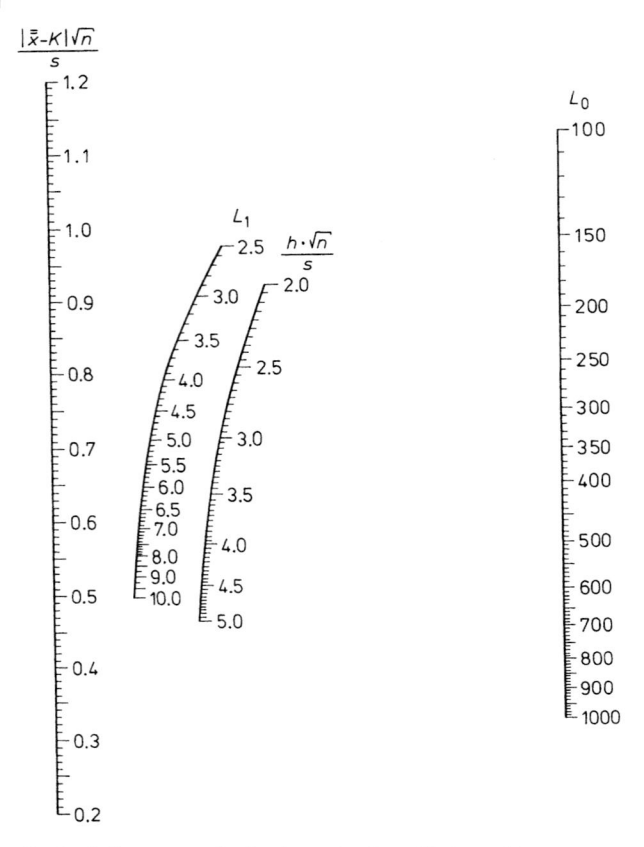

Fig. 2-24 Nomogram for the determination of h, L_0, and L_1.

The nomogram may also be used for the determination of the ARL (L_0 or L_1) if h is known. However, it should be kept in mind that in each case only one ARL can be read for a deviation in one direction. If the mean is monitored from both sides, as described above, the common ARL [202, 210, 216] results:

$$\frac{1}{ARL_{common}} = \frac{1}{ARL_{lower}} + \frac{1}{ARL_{upper}} \tag{83}$$

It should be noted that the following relationship between the parameters of the V-mask (d, θ) and the parameters of the numerical decision criteria (K, h) exists [139, 216]:

$$h = \frac{2 \cdot s}{\sqrt{n}} \cdot d \cdot \tan \theta \tag{84}$$

(for $w = 2s$)

$$K = \bar{x} + \frac{2 \cdot s}{\sqrt{n}} \cdot \tan \theta \tag{85}$$

Note: Since shifting of the V-mask can at any time provide a look at the past progress of the process, the V-mask technique is preferred over purely numerical decision criteria for unstable processes [139].

2.6.7.6.7 Advantages and Disadvantages of the Cusum Chart

2.6.7.6.7.1 Disadvantages of the Cusum Technique
1. The cusum control chart is more complicated than the $\bar{x}$-chart in the following aspects:
   - calculation of the cusum values,
   - scaling of the ordinate,
   - calculation of the parameters for determining out-of-control situations (V-mask, numerical decision interval).
2. The mean value used as the reference value k (V-mask) must be determined very exactly in the preliminary period.
3. Single deviations (gross errors) in the process are more apparent on the $\bar{x}$-chart.
4. The choice and calculation of decision criteria have not yet been sufficiently investigated for chemical analysis.
5. Gradual changes in the mean (increases and decreases within five sample intervals) are detected earlier on the $\bar{x}$-chart [141].

2.6.7.6.7.2 Advantages of the Cusum Technique
1. Even small changes in the mean cause a slope change of the cusum line, which in certain circumstances can be detected by purely visual methods.

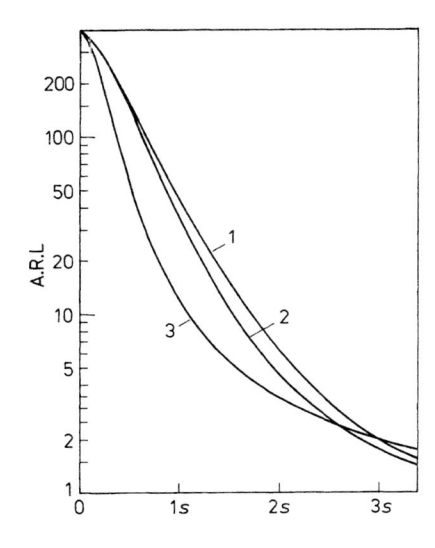

Fig. 2-25 Comparison of the ARL for:
1) $\bar{x}$-chart with 3s control limits,
2) $\bar{x}$-chart with 2s warning and 3s control limits,
3) cusum chart.

2. The cusum chart reacts more sensitively than the $\bar{x}$-chart to a shifting of the mean value in the range 0.5 to 2s [96, 109]; see Figure 2-25. As can be seen from Figure 2-26a, a change in mean from 0 to 0.2 does not signal an out-of-control situation on an $\bar{x}$-chart. On the other hand, the cusum line (see Figure 2-26b) clearly shows a change in slope after series number 30, i.e., the analysis process is "out of control".
3. The point in time at which the change in the process occurred is easily and relatively precisely determined.
4. The cusum chart is suited for monitoring processes with a high degree of imprecision.
5. In the application of numerical decision criteria, one may do without a graphical representation of the cusum values, although this prevents evaluation by a direct visual plausibility check.

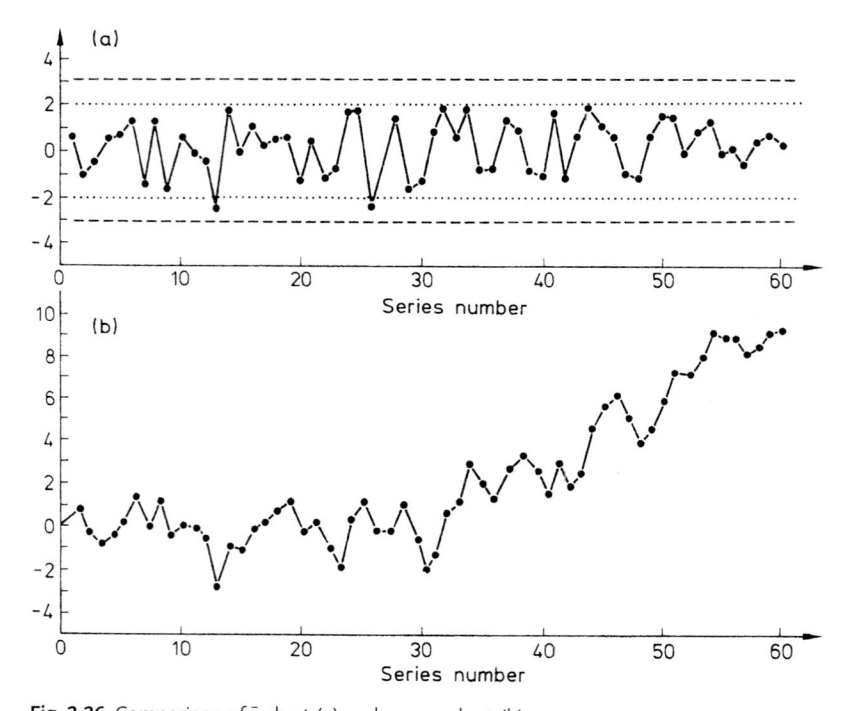

Fig. 2-26 Comparison of $\bar{x}$-chart (a) and cusum chart (b).

2.6.8
Summary of the Characterization of Control Charts

Table 2-9 succinctly characterizes frequently used control charts as well as their applicability.

2.6 The Control Chart System

Table 2-9 Overview of frequently used control charts.

Type of control chart	Calculations	Suitable for detecting	Required analyses
Mean value control chart	Mean value and standard deviation, control and warning limits in the preliminary period	Gross errors (deviation from the mean), systematic errors (trends, changes in the mean), random errors (scatter of individual values), generally for *precision control*	Control sample analyses at regular intervals
Difference chart	Mean difference, standard deviation, and control limits in the preliminary period; differences obtained from control sample analyses	Instrument drift	At least two determinations for each control sample at established intervals at the beginning and end of a series
R-chart (range chart)	Mean range, standard deviation, and control limits in the preliminary period; range from the control analyses	Imprecision	At least two determinations for actual samples at established intervals
Recovery rate (RR) chart	Mean RR, standard deviation, control and warning limits in the preliminary period; calculation of the RR from the results of the control analyses	Inaccuracy, proportional systematic errors	At least one analysis each of an unspiked and a spiked control or of an actual sample spiked with a known amount
Blank value control chart	Mean and standard deviation, control and warning limits in the preliminary period	Control of the blank values, errors in measuring instruments, reagent deviations	At least two blank value determinations at established intervals
Target value chart	Specified limits	Inaccuracy or imprecision	Depends on the required task
Cusum chart	Reference value and standard deviation from the preliminary period; setting the leading distance and half angles of the V-mask; calculation of the cumulative sum	Inaccuracy, drift	Control sample analyses at established intervals

3
Phase III: Routine Quality Assurance

3.1
Introduction

Quality assurance in routine analysis includes both internal laboratory quality control and all internal or external measures that are taken preventatively or as a result of alarming results. Routine refers here not only to the analyses, the quality of which should continue to be guaranteed, but also to the quality assurance measures that become a natural and routine part of the analysis.

3.1.1
Setting the Objectives of Phase III

Phase III quality assurance measures accompany routine analysis. They serve to maintain the reliability that was achieved in earlier phases and its continuing documentation. At the same time, this ensures comparability of analytical results and that they are legally binding.

3.1.2
Execution of Phase III

Internal quality assurance measures, both of a statistical and process technological nature, are the responsibility of every laboratory which performs analyses, whether on commission or within the framework of self-control. External quality assurance (Phase IV) can only take place in cooperation with the laboratories concerned. The preparation, organization, and evaluation of interlaboratory (or round robin) tests is the business of external control laboratories, which should above all assume an advisory role.

3.1.3
Progression of Phase III

Phase III may begin after completion of Phase II, the initial quality assurance phase.

Quality Assurance in Analytical Chemistry: Applications in Environmental, Food, and Materials Analysis, Biotechnology, and Medical Engineering, Second Edition. W. Funk, V. Dammann, G. Donnevert
Copyright © 2007 WILEY-VCH Verlag GmbH & Co. KGaA, Weinheim
ISBN: 978-3-527-31114-9

3 Phase III: Routine Quality Assurance

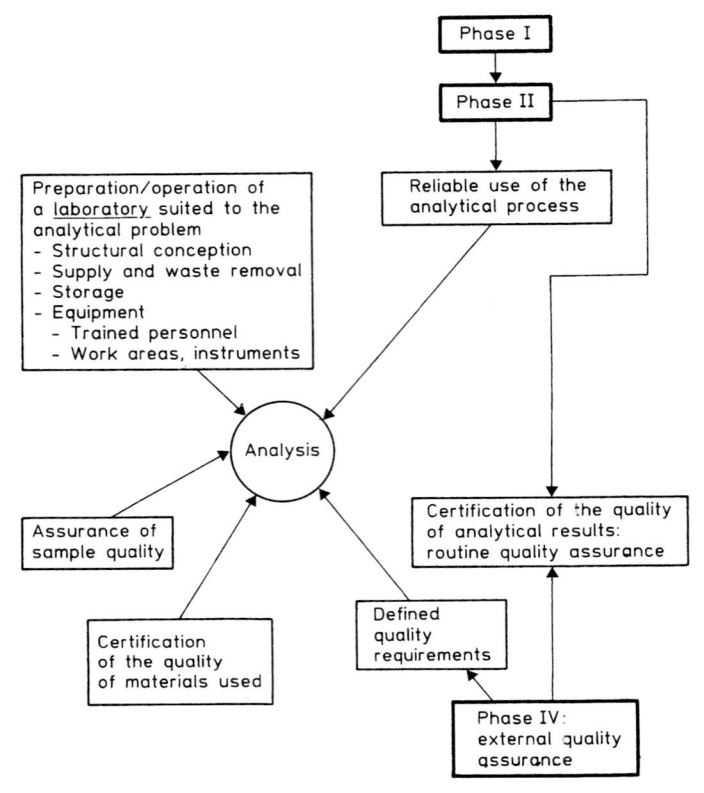

Scheme 3-1 Progression of Phase III, diagram.

While for each individual analytical process Phase II measures may proceed in isolation, Phase III must also include the environment in which the process takes place (Scheme 3-1).

According to supranational directives [165, 213], national, and international standards (ISO/IEC 17025 [76], ISO 9000 series [66–68], and EN 45000 series [60–62]), a defined analytical quality must be achieved, maintained, and proven by documentation. The elements of the quality assurance system extend from quality policy defined by the top management level down to the simplest operations at each workplace.

Following the ISO standards, the quality assurance system of an analytical laboratory should consist of

- the responsibility and accountability of the management,
- the principles of the quality system, including
  - the structure and responsibility of laboratory management,
  - an internal quality auditing system,
  - quality-related cost considerations,

- quality in procurement, purchasing,
  - quality of contracts with clients and subcontractors,
  - task management, scheduling and tracing of tasks and contracts,
- quality of the analytical process, including
  - quality of the laboratory, installations, and equipment,
  - material control and traceability,
  - validation of analytical methods,
  - quality control of the analytical process,
  - quality control of the analytical results,
  - quality and quality control of reporting and archiving results,
- control of measuring and test equipment, inspections, and calibration of instruments,
- maintenance,
- handling and disposal of wastes; environmental protection,
- handling of internal analytical quality problems,
- handling of externally indicated analytical quality problems,
- handling of customer complaints,
- corrective actions,
- quality records and documentation,
- personnel qualification and training,
- confidentiality and security (professional secrecy),
- use of statistical methods.

All topics mentioned above must be elaborated in a structured quality manual [129] of the laboratory. All related actions and procedures have to be described as a so-called standard operating procedure (SOP). When a given acknowledged standard is followed exactly, the description may be reduced to the citation of this standard. In all other cases, the operating procedures must be described in detail. The quality manual should be maintained and up-to-date.

Quality assurance in Phase III includes all the components of analysis:

- the laboratory from the perspective of planning and equipping,
- the equipment,
- the personnel,
- the analytical process:
  - spectrum of the process,
  - requirements (including quality assurance requirements),
- the samples,
- the reagents,
- the "outcome", quality control and documentation,
- the inclusion in an external quality assurance system.

No detailed chronological progression can be formulated for Phase III measures. Rather, the components of analysis in Phase III are subject to their own quality assurance systems, which, as far as statistical quality assurance is concerned, represent only an aid for the control of results.

3.2
Fundamental Measures of Internal Quality Assurance

3.2.1
The Laboratory and Laboratory Management

Reliable high precision analysis, especially in trace regions, requires a laboratory which is staffed, equipped, and managed accordingly. Quality assurance begins already in the planning and construction stages. In addition to compliance with pertinent construction codes, fire prevention regulations, and environmental and occupational safety codes, steps which will have a decisive influence on later analytical quality may be taken at this stage, if the future analysis operation is kept in mind:

- What path will the sample follow within the laboratory:
  - Where and how does it enter the laboratory?
  - Where and how will it be stored?
  - Does the transport path require subjecting the sample to unnecessary warming and recooling (i.e., interruption of the cooling chain)?
  - Is it possible that long transportation distances would prompt personnel to develop makeshift interim storage in the laboratory, which does not meet sample conservation requirements?
  - Is it certain that samples for trace analysis will not be contaminated on their way through the laboratory (e.g., by gases and vapors from the air, etc.)?

- Is the climate (humidity, heating, and especially ventilation) suitable for the planned analysis or can future contamination of trace analysis through the air not be ruled out?

- Have sufficient storage facilities been planned for samples, reagents, glass and plastic containers, and other material to be used?

- How have the progress of the operation and the paths of the personnel within the laboratory complex been conceived? Have spatial or operational bottlenecks been preprogrammed? Is it certain that rooms in which trace analysis is performed will not be contaminated by normal human and material traffic?

- Has future data networking been considered in planning the electrical system?

Errors in laboratory planning may appear during future laboratory operation as serious hindrances. Long, inconvenient transportation paths through the laboratory encourage makeshift arrangements, which stand in the way of quality assurance. Spatial restrictions and the crossing of personnel or material traffic can easily lead to unpredictable situations or even accidents under hectic working conditions. Personnel irritated by poorly organized laboratory operations are more prone to careless errors and discovering the cause of individual quality problems is accordingly more difficult.

3.2.2
Personnel

The personnel requirements of an analytical laboratory are almost a foregone conclusion:

- Sufficient personnel must be available.
- Personnel must be technically qualified and competent. This is not to be understood merely in terms of formal qualifications, but also in terms of a certain routine practice for obtaining reliable analytical results. This is something which specialists, who perform analyses only sporadically, or even trained temporary workers cannot possess.
- Quality assurance measures, training and practice phases, and continuing education must of course be taken into account in personnel capacity planning.
- The personnel, from management to temporaries, should be conscious of the need for responsibility and quality. Quality assurance systems always harbor the danger of being converted into personnel control systems. Quality *checks* should be performed by each worker in his or her area within the framework of a quality *assurance* system for which he or she is jointly responsible. The working atmosphere should not hinder open and honest discussion of quality problems.

3.2.3
Outfitting and Equipment

It should be obvious that from the standpoint of quality assurance the workplace must be equipped with state-of-the-art technology, and that up-to-date safety regulations, recognized international standards (e.g., bearing the CE mark in Europe), and ergonomics should be observed.

3.2.3.1 Performance Monitoring, Calibration and Adjustment of Measuring Equipment

Measuring and metering equipment, such as pipettes, thermometers, and balances, are subject to regular performance monitoring, also known as official verification or calibration [84].

The goal of the testing is to ensure the accuracy and stability of the measuring equipment. Their calibrations are to be executed at fixed and recurring intervals by accredited calibration laboratories. Quality assurance in the laboratory is thus the rigid adherence to the aforementioned periods for all tests. All calibrations should be documented ("calibration certificate").

Independent of the calibration is the *adjustment* of the analytical equipment, which serves as the validation and, if necessary, the readjustment of electronic circuitry ("equipment alignment"). Adjustment is carried out regularly at intervals recommended by the equipment manufacturer (see Table 3-1). Every adjustment must be recorded in the laboratory notebook including the date and related details.

3.2.3.2 Maintenance of Equipment

In addition to repairing defects, maintenance of equipment also means regular inspections and maintenance work [32]. Inspections and preventive maintenance are carried out periodically according to the recommendations of the manufacturer or based on practical experience (see Table 3-2). For example, maintenance intervals for photometers range from daily to six-monthly.

Table 3-1 Examples of instrument adjustments [5].

Instrument	Adjustment	Other aids
pH meter	hourly or daily	buffers of pH 4, 7, and 10
UV/Vis spectrophotometer	monthly	absorption; e.g., sulfuric acid/potassium dichromate solution and certified standards
IR spectrometer	quarterly	polyethylene standard
Analytical balances	half-yearly	test weights
Refractometer	half-yearly	glass test pieces
Spectrofluorimeter	half-yearly	fluorescence standards
Gas chromatograph	dependent upon column usage	comparison of the solubility of test samples

Table 3-2 Regular photometer maintenance and inspections.

Daily	
– before beginning measurements	If possible align the photometer (and recorder and printer) before each series
– after completing all measurements	Clean cuvettes, cuvette carriage, and cuvette holders
Weekly	Clean removable filters
Monthly	Replace tubes for flow-through cuvettes, clean dust filters and replace if necessary, check photometer bulb and multiplier, clean the outside carefully with a soft cloth
Half-yearly (company service)	Clean lens system, remove dust from inside of housing

Maintenance and repair work are to be recorded in the laboratory notebook (e.g., instrument data recording sheet according to GLP).

3.2.4
Materials

Materials used in analysis include solvents, reagents, disposable plastic and/or glass articles, standards, quality control samples, and the samples to be analyzed.

3.2.4.1 Certifying Sample Quality

In order to safeguard quality, only those samples for which the origin and "biography" are known and documented should be analyzed. The following information should be documented for each individual sample:

- when it was obtained,
- where (exactly) it was obtained,
- how it was obtained,
- who obtained it,
- how it was handled at the sample source,
- what it was transported in,
- how it was transported,
- where it was stored,
- for how long it was stored,
- how it was preserved,
- if and how it was homogenized before further treatment,
- how it was divided in the laboratory for analysis,
- and, in particular, how a positive identification can be ensured.

Depending on the purpose of the analysis, the source of the sample, and the analytical procedure used, ordinances, regulations, and guidelines are already in place and standards have been developed that unequivocally specify or regulate the taking, storage, and conservation of samples. Adherence to these rules must be guaranteed [35, 37–41, 73, 137, 182]. It should be kept in mind when taking the sample that the amount or volume of sample is large enough to perform at least one additional control analysis if necessary.

3.2.4.2 Analysis-Related Materials

Analysis-related materials include standard substances, solvents (including water) of varying grades of purity, other reagents, technical gases, sample receptacles, and disposable articles made of plastic and/or glass. All materials directly or indirectly related to the analysis must themselves fulfill certain quality requirements. This is valid for both the purity of chemicals as well as for materials, workmanship, and use of and/or (pre)treatment of containers, compounds, etc. (see also Section 3.5.1 "Sources of error in the analytical laboratory").

Before using chemicals or materials from a new batch for the first time, especially if the manufacturer has obviously changed or if brochures or accompanying documents indicate changes in production, they should be subjected to random quality control checks. For procedures requiring calibration, this means the performance of a completely new calibration with the determination and verification of process data (compare with Phase II: "Training" phase, Section 2.3).

Appropriate storage and usage of materials while observing occupational safety regulations [8] should be a matter of course. Disposable articles should only be used once and then disposed of professionally. Attention must also be paid to expiry dates.

3.2.4.3 Control Samples for Routine Quality Control

When routine quality control is maintained by means of control charts, the control sample batch should be available in sufficient quantity and have a sufficiently long shelf-life so that it may be used over a longer period of time (ideally one or more years).

If a new control sample batch is introduced, one must not forget to begin a new preliminary period immediately.

3.2.5
Instituted Analytical Processes

Only those analytical processes that are appropriate with respect to the accuracy and precision required for the analytical purpose and of which the reliability has been proven within the framework of a preliminary period (see also Section 2.2 and following sections) are suitable for use in routine analysis.

The selection of the analytical process to be implemented must be oriented around legal requirements and/or the state of technology, as documented in standard procedures that are relevant to the analytical problem. Use of modified or completely different analytical processes must be agreed upon with the client, and documentation is absolutely necessary.

An analytical process suitable for answering a certain question is not always the "best" procedure from a scientific point of view [110] (also see Chapter 0, Introduction). However, it is important that the results obtained are legally binding, i.e. that compliance with legal requirements [75, 103, 186] can be defended in court if necessary.

Substituting an established analytical process with one of greater inaccuracy is not acceptable; results obtained with different analytical processes, the equivalence of which has not been proven, *are not comparable and in most cases will not be accepted.*

3.2.6
Testing the Equivalency of Analytical Results

Procedures for testing the equivalency are described in the new version of DIN 38402-71 [48] as well as in ISO/TS 16489 [134]. Common to all of them is that first an equivalency test is carried out with matrix-free samples, i.e. standard solutions.

In the following, the subscript "R" indicates data that have been determined by a reference procedure. "CP" refers to data of the comparison procedure, which should be tested for equivalency against the reference procedure.

One has to differentiate between the detection of the *equivalency of analysis results from single-matrix samples* and the detection of the *equivalency of analysis results from multiple-matrix samples*.

3.2.6.1 Testing the Equivalency for a Single Matrix

A typical application of this type of equivalency testing is the use of rapid test methods for wastewater purification plants within the framework of self-control. With a uniform sample composition, one can verify whether the results of rapid tests are comparable to those of the reference procedure with a relatively small expenditure of effort.

3.2.6.1.1 Comparing the Scatter in Matrix-Free Solutions

Initially, the scatter is compared with respect to standard solutions. The F-tests in each case are then only carried out if the variation coefficient of the comparison procedure is greater than that of the reference procedure. If it is smaller, tests on natural samples are conducted immediately instead.

a) With procedures that can be calibrated, the variation coefficients (V_{xo}) of the calibration curves of the reference procedure (R) and comparison procedure (CP) are compared by means of an F-test. At this point, existing data can be referred to; the calibration range must not necessarily be identical, but merely similar.

$$F\text{-test: } TV = \frac{(V_{xoCP})^2}{(V_{xoR})^2} \tag{86}$$

Table values: $F(f_R = f_{CP} = N - 2; P = 99\%)$

Decision: If $TV \leq$ the F-value in the table, then the scatter is equivalent and the equivalency test can be continued with natural samples.

b) With procedures that cannot be calibrated, the variation coefficients (V_x) of the analytical results from at least five repeated determinations of standard solutions are compared by means of an F-test.

$$F\text{-Test: } TV = \frac{(V_{CP})^2}{(V_R)^2} \tag{87}$$

Table values: $F(f_R = f_{CP} = N - 2; P = 99\%)$

Decision: If $TV \leq$ the F-value in the table, then there is no significant difference in the scatter and the equivalency test can be continued with natural samples.

3.2.6.1.2 Equivalency Tests Using Natural Samples

The methodology for procedures that can and cannot be calibrated is the same. As mentioned before, the test is only carried out if the variance of the comparison procedure is larger than that of the reference procedure.

Aliquots from representative samples are each analyzed at least six times with the reference and comparison procedures. Each analysis series is first investigated for outliers by means of a Grubbs test, whereby a maximum of one outlier is

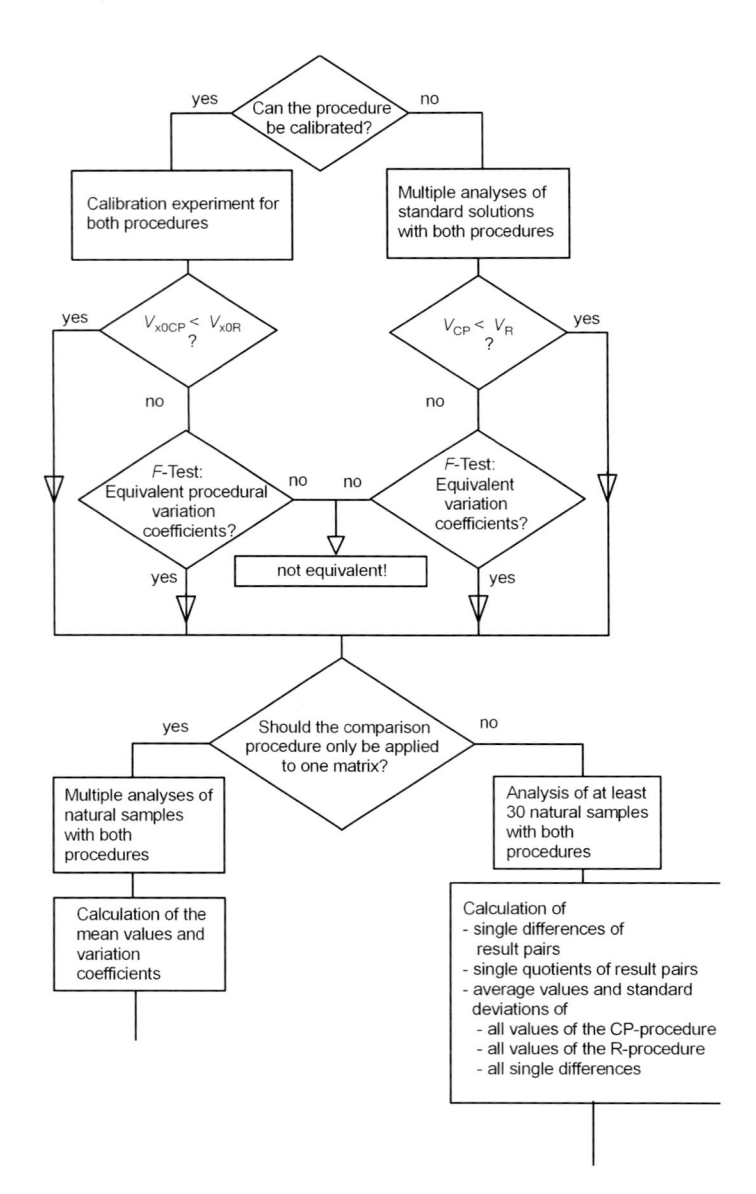

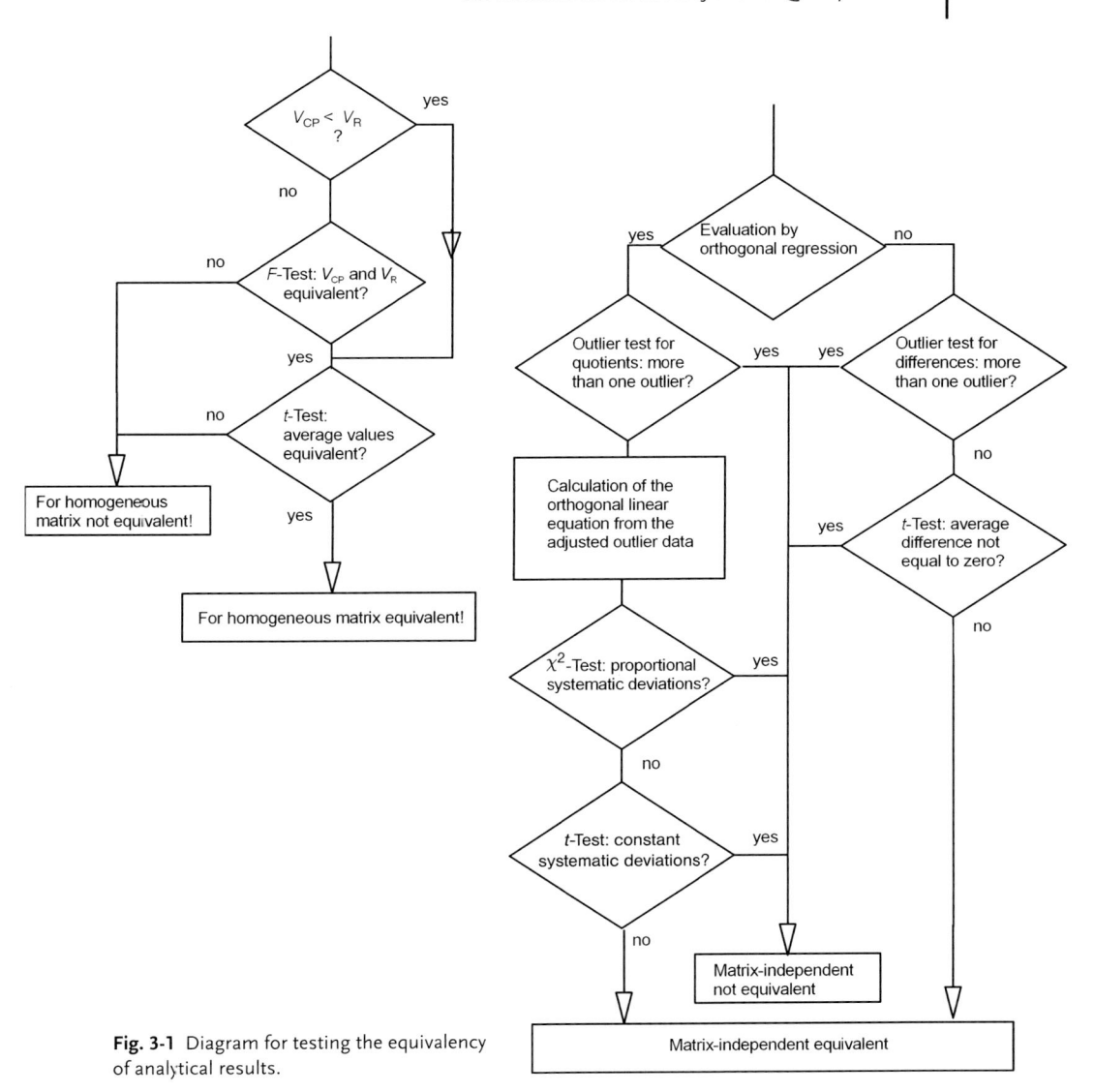

Fig. 3-1 Diagram for testing the equivalency of analytical results.

allowed in each series. Finally, the test of the equivalency of the variation coefficients is carried out:

$$F\text{-test: } TV = \frac{(V_{CP})^2}{(V_R)^2} \tag{88}$$

Table values: $F(f_R = f_{CP} = N-1; P = 99\%)$

Decision: If $TV \leq$ the table value, then the scatter is equivalent.

If the variation coefficient of the comparison procedure is not significantly larger than that of the reference procedure, the mean values are checked for significant differences by a mean *t*-test:

$$TV = \frac{|\bar{x}_R - \bar{x}_{CP}|}{s_d} \cdot \sqrt{\frac{N_R \cdot N_{CP}}{N_R + N_{CP}}} \qquad (89)$$

with

$$s_d = \sqrt{\frac{(N_R - 1) \cdot s_R^2 + (N_{CP} - 1) \cdot s_{CP}^2}{N_R + N_{CP} - 2}} \qquad (90)$$

Table values: $t\ (f = N_R + N_{CP} - 2;\ P = 99\%)$

Decision: If $TV \le$ the table value, then the results are equivalent.

If future analytical results over a wide substance-content interval (working range) are expected, the equivalency test must at least be carried out in the middle and at both ends of the range.

For non-equivalency of the tested comparison procedure to the reference procedure, it can be checked whether, in view of the quality demands in terms of precision and/or trueness of the intended application (e.g., in monitoring of drinking water), the reference procedure may have exceeded requirements. If these quality requirements are fully met by a non-equivalent comparison procedure, then this may nevertheless be suitable for the purpose.

3.2.6.2 Testing the Equivalency in Different Matrices

3.2.6.2.1 Comparing the Scatter in Matrix-Free Solutions

As described in Section 3.2.6.1.1, for procedures that can be calibrated, an initial comparison of the variation coefficient of the calibration slopes V_{xoCP} and V_{xoR} is carried out. If the procedure cannot be calibrated, then the variation coefficients V_{CP} and V_R, obtained from repeated determinations of standard solutions, should be compared.

If there is no significant difference between the variation coefficients in the matrix-free solutions, further tests can be carried out based on either *orthogonal regression* or the *difference method*.

3.2.6.2.2 Methodology of Orthogonal Regression

For this test, the following conditions must be met:

- At least $N = 30$ natural samples of varying matrices and varying substance contents should each be analyzed using the reference and comparison procedures, so that 30 pairs of values are available.

- The highest analytical result of the reference procedure must be at least five times and at most 100 times larger than the lowest.

- If the highest result is smaller than five times the lowest, the difference method must be used; if it is more than 100 times larger, the comparison range must be divided.

- So that the sample concentrations are distributed as evenly as possible over the relevant concentration range, the comparison range must be split into at least five equidistant concentration classes. These classes are filled equally with pairs of values from the routine analysis, and in the order of the analysis values. When a class is full, any remaining pairs for this class are discarded.

- In each case, single or multiple determinations can be carried out, but the number of repeated determinations must be the same for both procedures. For multiple determinations, the two mean values are related as a pair of values.

  From each pair, a quotient Q is calculated:

$$Q_i = \frac{x_{CPi}}{x_{Ri}} \tag{91}$$

- By means of a Grubbs outlier test, a maximum of one pair of values may be eliminated as an outlier (outlier pair).

a) Determination of Outlier Pairs

Outlier pairs are determined from the respective analytical values by means of a Grubbs test of the quotients of the individual pairs of values (Q^* is a suspected outlier quotient):

$$TV = \frac{|Q^* - \overline{Q}|}{s_Q} \tag{92}$$

with

$$\overline{Q} = \frac{\sum Q_i}{N} \tag{93}$$

and

$$s_Q = \sqrt{\frac{\sum (Q_i - \overline{Q})^2}{N-1}} \tag{94}$$

The test value, TV, is compared with the table value, rM ($f = N$; $P = 95\%$). If TV is larger than the table value, the tested pair of values must be eliminated as an outlier. The mean and standard deviation are calculated anew from the remaining data set. Other suspected outlier single data are likewise checked according to the described procedure.

If such an outlier pair appears, it is acceptable to remove this pair from the orthogonal regression (maximum one pair). However, this pair of values is to be mentioned in the documentation and classified as an outlier pair. Conditions (e.g., matrix type, possible causes of the deviation, etc.) must also be documented and commented upon.

b) Calculation of Linear Equations

In contrast to simple linear regression, in which only the y-values are considered to be affected by errors, both x- and y-values can be regarded as erroneous in the case of orthogonal regression. Accordingly, the slope and axis intercept are calculated using different formulae. Ideally, orthogonal regression yields a line with a slope of $b = 1$ and an intercept of $a = 0$.

In the equivalency test, the results of the reference procedure are plotted on the abscissa, those of the comparison procedure on the ordinate.

The linear equation: $y = a + b \cdot x$ thus becomes $x_{CP} = a + b \cdot x_R$

Slope:

$$b = \frac{s_{CP}}{s_R} \qquad (95)$$

with

$$s_{CP} = \sqrt{\frac{1}{N-1} \sum (x_{CPi} - \bar{x}_{CP})^2} \qquad (96)$$

and

$$s_R = \sqrt{\frac{1}{N-1} \sum (x_{Ri} - \bar{x}_R)^2} \qquad (97)$$

Axis intercept:

$$a = \bar{x}_{CP} - b \cdot \bar{x}_R \qquad (98)$$

c) Testing for Proportional Systematic Deviation: χ^2-Test

Test value:

$$x^2 = N \cdot \ln\left(\frac{s^4 - s_{RCP}^4}{s_R^2 s_{CP}^2 - s_{RCP}^4}\right) \qquad (99)$$

with

$$s = \sqrt{\frac{1}{2} \cdot (s_R^2 + s_{CP}^2)} \qquad (100)$$

and

$$s_{RCP} = \sqrt{\frac{1}{N-1} \sum (x_{Ri} - \bar{x}_R)(x_{CPi} - \bar{x}_{CP})} \qquad (101)$$

If $x^2 \leq 3.8$, then no proportional systematic deviation is present with an error probability of 5%.

d) Testing for Constant Systematic Deviation

If a constant systematic deviation is present, the regression line is shifted parallel in relation to the bisector. The shift corresponds to the difference in the mean values (bias):

Shift:

$$\bar{D} = \bar{x}_{CP} - \bar{x}_R \tag{102}$$

The presence of a constant systematic deviation can be tested by means of a joint t-test.

Joint t-test:

$$TV = \frac{|\bar{x}_R - \bar{x}_{CP}|}{s_D} \sqrt{N} \tag{103}$$

with

$$s_D = \sqrt{\frac{1}{N-1} \sum (D_i - \bar{D})^2} \tag{104}$$

If $TV \leq t\,(f = N-1;\ P = 99\%)$, no constant systematic deviation is present.

3.2.6.2.3 Methodology of the Difference Method

For this test, the following conditions must be met:

- At least $N = 30$ pairs of values.
- The lowest and the highest final result of the reference procedure may differ by at most a factor of five.
- There is no organization into concentration classes.
- In each case, single or multiple determinations can be carried out, but the number of repeated determinations must be the same for both procedures. For multiple determinations, the two mean values are related as a pair of values. The difference is calculated for each pair:

Single difference:

$$D_i = x_{Ri} - x_{CPi} \tag{105}$$

- A maximum of one outlier pair can be eliminated.

a) Determination of Outlier Pairs

Outlier pairs are determined from the respective analytical values by means of a Grubbs test of the differences D_i of individual pairs of values (D^* is a suspected outlier difference).

Mean difference:

$$\bar{D} = \frac{\sum D_i}{N} \quad (106)$$

Grubbs test:

$$TV = \frac{|D^* - \bar{D}|}{s_D} \quad (107)$$

with

$$s_D = \sqrt{\frac{\sum (D_i - \bar{D})^2}{N - 1}} \quad (108)$$

The test value, TV, is compared with the table value, rM ($f = N$; $P = 95\%$). If TV is larger than the table value, the tested difference must be eliminated as an outlier. The mean and standard deviation are calculated anew from the remaining data set. Other suspected outlier single data are likewise checked according to the described procedure. At most one pair of values can be removed from the analysis according to the difference method. However, this pair is to be mentioned in the documentation and classified as an outlier pair. Conditions (e.g., matrix type, possible causes of deviation, etc.) must also be documented and commented upon.

b) Detection of Equivalency by a Joint *t*-Test

For the equivalency test, the test value should be calculated according to the following equation:

$$TV = \frac{|\bar{D}|}{s_D} \cdot \sqrt{N} \quad (109)$$

If $TV \leq t$ ($f = N - 1$; $P = 99\%$), there is no significant difference. The results are equivalent.

Please note: A *general* statement about the equivalency of two procedures is *not* possible. Therefore, "equivalency" is defined in a more conventional sense as *"fulfilling the standard"*.

3.2.7
Uncertainty of Measurements

Some years ago, with the advent of the term "uncertainty in measurement", international standardization acquired a new concept for accuracy or lack thereof. Analytical results are to include their measurement uncertainty. ISO/IEC 17025 requires test laboratories to have procedures in place for the estimation of the uncertainty in measurements and to apply them. A set method of estimation is not specified, though a "reasonable" estimation of the measurement uncertainty should be conducted, one which is based on knowledge of the implementation of the procedure and the type of measurement, as well as, for example, on previous experience and the use of validation data.

The fundamental document for the estimation of uncertainty in measurement is the *ISO Guide to the Expression of Uncertainty in Measurement, GUM*, which is also available as ENV 13005 [75]. In Europe, however, the EURACHEM/CITAC guide *Quantifying Uncertainty in Analytical Measurement* for use in analytical chemistry is available in place of this very formal and mathematical manual [90].

Beyond that, there is a series of publications [7, 93, 111, 135] in which various models for the estimation of uncertainty in measurement are described, though (at the time of this writing, 2006) there is still no agreed concept.

Common to all of these concepts is the principle of considering all the important sources of error that can falsify (or rather, make uncertain) an analytical result. The total uncertainty, otherwise known as the measurement uncertainty, is to be calculated. The final analytical result is indicated with ± uncertainty in measurement; the interval corresponds to a confidence interval in which the true analytical result lies with a specific certainty (significance level). The significance level is given alongside the result uncertainty.

3.2.7.1 New Terms According to the EURACHEM Guide [90]

Preliminary Remarks

- Influencing variables that cause uncertainty in measurement are denoted by "X". These include, for example:
  - the volume measurement,
  - the temperature dependency of volumes, reactions, etc.,
  - the weighing,
  - the location (different laboratories), and
  - the time period.

  These last two constitute parts of the so-called repeatable or comparable conditions and are documented in interlaboratory test data.

- with "x" as the estimated values of "X",
- "Y" is the measurement being determined,
- "y" is the estimated value of "Y",
- "u" is a (standard) deviation, the uncertainty in measurement or analytical results.
- "U" is the uncertainty in measurement as half the width of the confidence interval.

Error (or Measurement Deviation)
The difference between an individual result and the true value of the measurement. The error cannot be exactly determined as the true value is unknown.

Random Measurement Deviation
The proportion of the measurement deviation that is due to unpredictable fluctuations of the influencing variables and can be reduced by increasing the number of determinations.

Systematic Measurement Deviation
Components of the deviation that remain constant through multiple determinations of the measurement (constant systematic deviation) or change in predictable ways (proportional systematic deviation). The measurement result should be corrected for all known systematic effects, even though there remains a proportion of uncertainty of the systematic deviation to take into account.

Measurement Uncertainty
One of the parameters assigned to the measurement result that indicates the spread of the values that can be reasonably assigned to the measured variable. These parameters can, for example, be a standard deviation or another part of a range that specifies a particular confidence level.

Standard Uncertainty $u(x_i)$
The components of uncertainty expressed as a standard deviation, whereby x_i is the estimated value for the input parameter X_i.

Combined Uncertainty $u_c(y)$
The total uncertainty resulting from all variance components according to the law of error propagation; here, y is the measurement result for the measured quantity Y.

$$u_c(y) = \sqrt{\sum (u(x_i))^2} \tag{110}$$

Expanded Uncertainty $U(y)$
This indicates the value range that contains the true value of the measured quantity with high probability, P:

$$U(y) = k \cdot u_c(y) \tag{111}$$

Here, k is the coverage factor

$k = 2$ corresponds approximately to the significance level $P = 95\%$

Please note: Strictly speaking, the factor $k = 2$ applies only to $f \geq 30$ degrees of freedom. If the estimation of the uncertainty in measurement is based on a smaller number of analytical data, the corresponding value t (f, $P = 95\%$) from the t-table should be entered for the value of k [94].

The value range for the true value of measured quantity Y takes the form:

$y \pm U(y)$

3.2.7.2 Overview of Common Procedures for the Determination of Measurement Uncertainty

For a better understanding, the different concepts are applied to a fictitious analytical example (for the data, see Table 3-3).

Table 3-3 Fictitious data for the determination of uncertainty in measurement with different procedures.

Sequence i	Description	Actual value	Standard uncertainty	Relative standard uncertainty, u_i
1	Substance to be analyzed, ABC = measured quantity, Y actual analytical result, y	467.2 mg/l	to be determined	to be determined
2	Analytical procedure	XYZ		
3	Working range	100 to 900 mg/l		in the middle
	Uncertainty of standard solutions		1.5 mg/l	0.27%
4	Number of standards used during calibration, N_c	10		
	Number of analyses for each sample, N_a	1		
5	Calibration:			
5.1	Process standard deviation and variation coefficient of the procedure, V_{xo}		3.5 mg/l	0.63%
	Formula for the standard uncertainty of the analytical result y, which can be calculated from the calibration function: $$u(y) = s_{xo} \cdot \sqrt{\frac{1}{10} + 1 + \frac{(y-550)^2}{825000}}$$		3.67 mg/l	
	$$u(y)_{rel} = \frac{u(y)}{y} \cdot 100\%$$ for y = 467.2 mg/l (out of 1)			0.79%
6	The analysis involves five steps:			
6.1	Measure the analytical sample	5 ml	0.01 ml	0.2%
6.2	Dilute the sample			0.2%
6.3	Extract ABC			2.9%
6.4	Add reagent			0.2%
6.5	Measure the sample:			
6.5.1	Measurement uncertainty of the measuring equipment			0.1%
6.5.2	Daily calibration of the measuring equipment with a control solution; uncertainty of the target value			0.4%

Table 3-3 (continued)

Sequence i	Description	Actual value	Standard uncertainty	Relative standard uncertainty, u_i
7	Participation in **interlaboratory tests** for proficiency testing. Data from the interlaboratory tests:			
7.1	Aggregated mean value from $n = 12$ participating laboratories	531.2 mg/l		
7.2	Repeatability standard deviation, s_r; same value as used in the method validation		5.9 mg/l	1.1%
7.3	Laboratory mean value; laboratory bias = (542.1 − 531.2) mg/l	542.1 mg/l	10.9 mg/l	2.1%
7.4	Reproducibility standard deviation, s_R		18.1 mg/l	3.4%
7.5	Standard deviation between laboratories, s_L $$s_L = \sqrt{s_R^2 - s_r^2}$$		9.2 mg/l	3.2%
8	Laboratory-internal quality assurance: **data from quality control charts**			
8.1	Average recovery rate; laboratory-internal proportional systematic deviation	101.3%		1.3%
8.2	Mean value control chart: mean and standard deviation	475.0 mg/l	3.8 mg/l	0.8%
8.3	Grand average determined for the analytical results during the control period using the range control charts	427.3 mg/l		
	Range control charts: calculated standard deviation		4.1 mg/l	1.0%
8.4	Analysis of certified reference material (c_{CRM} = 513 mg/l) total mean value $\bar{c}$ of all analytical results ($n = 30$), determined during the control period	508 mg/l	2.5 mg/l	0.5%
	Standard deviation, s		7.5 mg/l	1.5%
	Average recovery rate, $\bar{R}_m$	0.99		

3.2.7.2.1 Fundamental Concept of Combined Measurement Uncertainty

An analytical result is the consequence of several essential actions. These comprise not only all the steps of the analytical process itself, but also the quality assurance procedures, such as the use of certified reference materials for the adjustment of measuring equipment, the analysis of control samples for the operation of quality control charts, participation in interlaboratory tests, and so on. With regard to error propagation, each action carries its own uncertainty into the total measurement uncertainty of the analytical result. For simplification, the calculation is always carried out with the relative (i.e., percentage) uncertainties.

3.2.7.2.2 Concept of the EURACHEM Guide

The determination of measurement uncertainty follows a four-step concept [90]:

- Step 1: Specification of measured quantity Y
  For example, from Table 3.3: ABC

- Step 2: Identification of uncertainty sources x_i ("i" from Table 3.3)

$i = 3$:	Standard solutions for the calibration
$i = 5.1$:	Calibration
$i = 6.1$:	Measurement of the analytical sample
$i = 6.2$:	Sample dilution
$i = 6.3$:	Extraction
$i = 6.4$:	Reagent volume
$i = 6.5.1$:	Measuring equipment
$i = 6.5.2$:	Calibration of measuring equipment
$i = 7.2$:	Repeatability from a validation interlaboratory test

- Step 3: Quantification of the uncertainty components $u(x_i)$

$i = 3$:	Standard solutions for the calibration: $u_3 = u(CS)_{rel} = 0.27\%$
$i = 5.1$:	Calibration: $u_{5.1} = u(C)_{rel} = 0.79\%$
$i = 6.1$:	Volume measurement of the analytical sample: $u_{6.1} = u(V)_{rel} = 0.2\%$
$i = 6.2$:	Sample dilution: $u_{6.2} = u(D)_{rel} = 0.2\%$
$i = 6.3$:	Extraction: $u_{6.3} = u(E)_{rel} = 2.9\%$
$i = 6.4$:	Reagent volume: $u_{6.4} = u(V)_{rel} = 0.2\%$
$i = 6.5.1$:	Measuring equipment: $u_{6.5.1} = u(ME)_{rel} = 0.1\%$
$i = 6.5.2$:	Calibration of measuring equipment: $u_{6.5.2} = u(C_{ME})_{rel} = 0.4\%$
$i = 7.2$:	Repeatability from a validation interlaboratory test: $u_{7.2} = u(s_r)_{rel} = 1.1\%$

Simplification by combining the sources that have already been covered in existing data:

- for $i = 6.1$: The uncertainty of the pipette and the effect of the temperature were already summed during measurement of the analytical sample
- for $i = 6.2$: The uncertainty of the pipette and the effect of the temperature were already summed during dilution of the sample
- for $i = 6.3$: All components of uncertainty, such as quantity, purity of the extracting agent, effect of the temperature, etc., were already summed during extraction
- for $i = 6.4$: The uncertainty of the pipette and the effect of the temperature were already summed in the reagent volume
- for $i = 6.5.1$: The manufacturer of the measuring equipment has already noted, summarized, and documented all components of uncertainty in the equipment specifications. A prerequisite for the application of these specifications is a functioning monitoring system for the measuring and testing equipment.
- for $i = 6.5$: Errors relating to the measuring equipment can additionally be combined as follows:

$$u_{6.5} = \sqrt{u_{6.5.1}^2 + u_{6.5.2}^2} = \sqrt{(0.1\%)^2 + (0.4\%)^2} = 0.41\%$$

- Step 4: Calculation of the combined uncertainty $u_c(y)$

$$u_c(y) = \sqrt{u_3^2 + u_{5.1}^2 + u_{6.1}^2 + u_{6.2}^2 + u_{6.3}^2 + u_{6.4}^2 + u_{6.5}^2 + u_{7.2}^2} =$$
$$= \sqrt{0.27^2 + 0.79^2 + 0.2^2 + 0.2^2 + 2.9^2 + 0.2^2 + 0.41^2 + 1.1^2}\% =$$
$$= 3,3\%$$

Calculation of the expanded uncertainty $U(y)$
$U(y) = k \cdot u_c = 2 \cdot 3.3\% = 6.6\%$

Thus, the analytical result is 467.2 mg/l $\pm$ 6.6%.

Detailed examples can be found in the appendix of ref. [90].

For the estimation of uncertainty in measurement with less expenditure of effort, a so-called top-down model has been developed, in which the calculation is performed on the basis of existing precision and validation data. These data are all based on grouping the components of uncertainty, whereby single components are unobservable but are part of an *experimentally* determined combined uncertainty.

One such model, for example, is a complete analytical process through which estimated values of repeatability, reproducibility, and trueness are determined in the framework of interlaboratory tests. The uncertainty in measurement is made up of the laboratory standard deviation and the systematic deviation of the laboratory ("laboratory bias").

Another model is similarly based on the laboratory bias determined in interlaboratory tests as well as the precision data from internal quality assurance.

Top-down models are described in ISO/TS 21748 [135], in the NORDTEST report TR 537 [111], in a VAM report [7], and in a Eurolab technical report [93].

3.2.7.2.3 Procedures According to ISO/TS 21748 [135]

The estimation of uncertainty in measurement is carried out according to the following scheme:

a) Collate estimated values of repeatability, reproducibility, and trueness of the procedure from published data, in particular from the respective interlaboratory tests in which one's own laboratory has participated.
b) Assess whether the laboratory bias (constant systematic deviation) lies within the expected range according to a).
c) Assess whether the precision attained in the laboratory lies within the expected range according to a).
d) Test whether further influential factors exist which have not yet been accounted for in published studies; estimate variances that might result.
e) If the bias and precision are under control, calculate the estimated value for the combined uncertainty from the available data.

The basic model for indicating results with an estimation of uncertainty in measurement according to ISO/TS 21748 has the form:

$$y = m \pm u(y)$$
$$= m \pm (B + e)$$

here
m is the expected value
B is the laboratory component of the bias
e denotes random errors under repeatable conditions

If the laboratory bias and the laboratory precision are not significantly different from those values obtained in interlaboratory tests and there are no further components of uncertainty to consider, the total standard deviation from the validating interlaboratory test can be consulted for an estimation of uncertainty in measurement.

Reason: In this case, the following applies:

$u(B) = s_L$ (s_L is the standard deviation between laboratories according to ISO 5725-2)

$u(e) = s_r$ (s_r is the repeatability standard deviation according to ISO 5725-2)

$$u_c = \sqrt{u(B)^2 + u(e)^2} = \sqrt{s_L^2 + s_r^2} \tag{112}$$

The square root $\sqrt{s_L^2 + s_r^2}$ corresponds to the formula for the calculation of the reproducibility standard deviation s_R according to ISO 5725-2.

For the fictitious example, this means:

7 Interlaboratory test characteristics:
    7.1 Grand average: 531.2 mg/l
        (if it were known, the conventional true value would be better)
    7.2 Repeatability standard deviation s_r: 5.9 mg/l corresponds to $s_{r\,rel} = 1.1\%$
    7.3 Laboratory mean value: 542.1 mg/l
    7.4 Reproducibility variation coefficient CV_R: 3.4%
    7.5 Standard deviation between laboratories s_L: 3.2%

8 Data from the internal laboratory quality assurance:
    8.1 Uncertainty from the range control chart s_{rel}:   1.0%

$$\text{Calculated laboratory bias: } u(B) = \frac{542.1\,\text{mg/l} - 531.2\,\text{mg/l}}{531.2\,\text{mg/l}} \cdot 100\% = 2.05\%$$

Findings:
(a) $u(B)$ is smaller than s_L but is of the same order of magnitude.
(b) s_{rel} is of the same order of magnitude as $s_{r\,rel}$.

Conclusion: The reproducibility standard deviation from the interlaboratory test may be directly used as the measurement uncertainty: $u(y) = s_{R\,rel} = 3.4\%$

Calculation of the expanded uncertainty: $U(y) = 2 \cdot 3.4\% = 6.8\%$

Indication of the analytical result: 467.2 mg/l ± 6.8%

3.2.7.2.4 Method According to the NORDTEST Report TR 537 [111]

The NORDTEST report describes procedures in which precision data and systematic components are combined for the estimation of the measurement uncertainty. One characteristic of the *precision* is the so-called uncertainty based on the reproducibility within the laboratories $u(R_W)$, which can be derived from control charts, for example.

The basic model for the indication of results with estimation of the measurement uncertainty takes the following form:

$$y = m \pm u(y)$$
$$= m \pm ((\delta + B) + e)$$

here
m is the expected value
δ is the method bias
B is the laboratory component of the bias
e is the allowed error under repeatable conditions

For the estimation of *systematic components*, there are three alternatives:
- the method bias $u(C_{ref})$, e.g. from the examination of reference material,
- the laboratory bias $u(B)$, e.g. from proficiency tests,
- systematic deviations from the results of recovery experiments.

The NORDTEST model is now illustrated with the established *fictitious example*:

1. *Specification of the measured quantity:* ABC

2. *Quantification of the reproducibility within the laboratory:*
   A: Mean value control chart ($c = 475$ mg/l) $\Rightarrow u_{8.2} = u(R_w) = s_{Rw\,rel} = 0.8\%$
   B: Are there further uncertainty components that have not been covered? No

3. *Quantification of systematic components:*
   Participation in the proficiency test: laboratory bias $u(B) = u_{7.3} = 2.1\%$
   (if the laboratory has already participated in several interlaboratory tests, calculate the laboratory bias as the square root of the mean value of the squares of the single bias values)
   Uncertainty of the assigned value: $u(C_{ref}) = s_R/\sqrt{n} = u_{7.4}/\sqrt{n} = 3.4\%/\sqrt{12} = 0.98\%$
   (when participating in more than one interlaboratory test, s_R and n are calculated as arithmetic mean values of these measurements from the proficiency tests)

4. *Conversion of the uncertainty components into standard uncertainty $u(x)$:*
   $u(R_W) = 0.8\%$
   $$u(bias) = \sqrt{u(B)^2 + u(C_{ref})^2} \tag{113}$$
   $$= \sqrt{2.1^2 + 0.98^2}\% = 2.32\%$$

5. *Calculation of the combined uncertainty*
   $$u_c = \sqrt{u(R_w)^2 + (u(bias))^2} \tag{114}$$
   $$= \sqrt{0.8^2 + 2.32^2} = 2.45\%$$

6. *Calculation of the expanded uncertainty*
   $U = 2 \cdot u_c = 2 \cdot 2.45 = 4.9\%$

Therefore, the measurement uncertainty for ABC in the relevant concentration range is indicated as $\pm 4.9\%$.

NORDTEST recommends the use of the interlaboratory test data from the respective previous three years along with the mean values of the laboratory bias, the reproducibility standard deviation s_R, and the number of participants n.

3.2.7.2.5 Procedures According to VAM

The estimation of the measurement uncertainty from repeated measurements of a representative certified reference material is described in ref. [7]. It is assumed that the reference material is routinely analyzed over a long period of time and is thereby subjected to the whole analytical method, so that all important uncertainty components are covered.

Here again, the combined uncertainty is made up of precision and trueness contributions:

$$u_c(y) = \sqrt{s_{rel}^2 + u(\bar{R}_m)^2} \tag{115}$$

For the example data (8.4), there is the following measurement uncertainty of the result 467.2 mg/l ABC:

- *Relative standard deviation of the results for the reference material:*
  $s_{rel} = 0.0148 \,(= 1.5\%)$

- *Relative uncertainty of the mean recovery rate of the reference material $u(\bar{R}_m)$:*

$$u(\bar{R}_m) = \bar{R}_m \cdot \sqrt{\left(\frac{s^2}{n \cdot \bar{c}^2}\right) + \left(\frac{u(c_{CRM})}{c_{CRM}}\right)^2} \tag{116}$$

$$u(\bar{R}_m) = 0.99 \cdot \sqrt{\left(\frac{7.5^2}{30 \cdot 508^2}\right) + \left(\frac{2.5}{513}\right)^2} = 0.0056$$

- *Combined uncertainty:*

$$u_c(ABC) = \sqrt{0.0148^2 + 0.0056^2} = 0.0158$$

- *Expanded uncertainty:*

$$U(ABC) = k \cdot u_c(ABC) = 2 \cdot 0.0158 = 0.0316 \approx 0.032 = 3.2\%$$

3.2.7.2.6 Simplified Methodology According to EUROLAB

The European Federation of National Associations of Measurement, Testing and Analytical Laboratories advocates a practice-oriented methodology [93]. As a basic rule for the estimation of measurement uncertainty, the expenditure of effort for the determination should be guided by the "fit-for-purpose" principle, that is to say, the measurement uncertainty indication should fulfill but not exaggerate the requirements. A laboratory with a good quality management system should therefore be in a position, if such a thing is stipulated, to give the measurement uncertainty with little expenditure of effort. Preferably, all relevant sources of uncertainty (e.g., sampling, sample transport, sample preparation, environmental conditions, personnel, and instruments) should be accounted for. For the calculation of the uncertainty, the available quality assurance data (typically, standard deviation) are used. These data must be representative of the respective measurement task.

Of the estimation methods that appear in the EUROLAB technical report, those that employ the total standard deviation taken from either a proficiency test or from the published characteristics of the standard procedure are particularly suitable. It is a prerequisite that the interlaboratory test samples are representative of the actual working range in terms of content and matrix.

3.2 Fundamental Measures of Internal Quality Assurance | 141

Example: Measurement uncertainty of the result 467.2 mg/l ABC

- Reproducibility standard deviation from the interlaboratory test $s_R = 3.4\%$ for all laboratories
- Quantification of the standard uncertainty:
  $s_R = u(all) = u_c = u_{7.4} = 3.4\% \Rightarrow 18.1$ mg/l for 531.2 mg/l
- Calculation of the expanded uncertainty:
  $U = k \cdot u_c = 2 \cdot 18.1$ mg/l $= 36.2$ mg/l

The laboratory must, however, be able to prove that it has either:

- participated in this interlaboratory test,
- that the following were successfully accomplished:
  - internal laboratory standard deviation was not significantly greater than s_r
  - no significant systematic deviation (bias) of its laboratory mean value was present compared to the grand average,
- and prove, with assistance from the mean value- or range- and the *RR*-control chart, that it has kept this quality consistent,
- or, by applying the data from the standard procedure, prove, with assistance from the mean value- or range- and the *RR*-control chart, that it has kept this quality consistent, and
- that it has participated in an appropriate external quality assurance.

It is worth noting that in place of the relative measurement uncertainties, the absolute values are taken here. Within the validated range this is reasonable. It should be kept in mind, however, that for the measurement indication of small measurements, the relative measurement uncertainties increase hyperbolically:

$$U_{rel} = \frac{U_{absolute}}{measurement} \cdot 100\%$$

The limit of quantification can be calculated very easily with $U_{absolute}$ from the maximum allowed measurement uncertainty U_{rel_max}:

$$x_{LQ} = \frac{U_{absolute}}{U_{rel\_max}} \cdot 100\%$$

Example: For a maximum permissible relative measurement uncertainty of 30% and a measurement uncertainty of 40 mg/l, the limit of quantification amounts to

$$x_{LQ} = \frac{40 \text{ mg/l}}{30\%} \cdot 100\% = 133 \text{ mg/l}$$

In ISO/TS 21748 [135], the NORDTEST handbook [111], and the VAM protocol [7], a number of methodologies are described that can be applied if the respective basic model is not applicable. No further details of these are given here.

3.2.7.2.7 Special Operational Procedures

In the framework of ISO 13530 [136] a special case is described for the estimation of measurement uncertainty, namely for operational (empirical) procedures such as sum parameters. With these procedures there are by definition no systematic deviations. Therefore, the estimation of the measurement uncertainty is based exclusively on precision data. The total standard deviation is given by:

$$s_t = \sqrt{\sum s_{x_i}^2} \tag{117}$$

with: s_{x_i} = single standard deviation

Here, very small single standard deviations – for example, those which amount to less than one-third of the maximum single standard deviation and hardly contribute to the total uncertainty – can be neglected, that is, omitted from the calculation.

3.2.7.2.8 Inclusion of the Pre-Laboratory Steps of the Analysis Method in the Calculation of the Measurement Uncertainty

For all of the models for calculating the measurement uncertainty presented thus far, it has been assumed that the sample is considered from its entry into the analytical procedure. Its previous history – sampling, transport, storage, and if necessary, homogenization and splitting – has been disregarded, just like in the case of interlaboratory test samples. However, if the laboratory is to supply not just an analytical result for a submitted sample, but a separate conclusion about the investigated item from which that sample was taken, the relative uncertainties of the preceding steps must also be quantified, documented, and with the help of the error propagation formula, included in the total uncertainty calculation:

$$u_{c\_total} = \sqrt{\sum u_{ci}^2} =$$
$$\sqrt{u_c^2(\text{sampling}) + u_c^2(\text{transport}) + u_c^2(\text{preservation}) + u_c^2(\text{storage}) + \cdots + u_c^2(\text{analysis})}$$

3.2.7.3 Indication of Measurement Uncertainty in Test Reports

According to ISO 17025, the measurement uncertainty should only be given if:

- it is important for the validity or application of the inspection results,
- the customer requires it,
- the compliance with limit values is in question.

It is necessary to indicate on what the estimation is based and it must be noted whether it concerns a simple standard deviation or an expanded measurement uncertainty.

Example:

SO_4^{2-} in waste water (IC method, ISO 10304-2):
  100 mg/l ± 12 mg/l*) or
  100 mg/l ± 12%*)

*) The measurement uncertainty was derived from the interlaboratory test data for the method validation. It represents an expanded uncertainty and was obtained through multiplication by the expansion factor $k = 2$; this corresponds to a confidence level of 95%.

3.2.7.4 Interpretation of Measurement Uncertainty in the Context of Limit Value Monitoring

The finding "upper limit value exceeded" (or "below the lower limit value") is affected by whether or not the measurement uncertainty has already been accounted for during the limit value determination. This is the case for the limit values according to waste water regulation [204], for example, though here only the applied analytical procedures are prescribed, and a given, required accuracy is missing from the limit value monitoring. In contrast to this, the EU Drinking Water Regulations [186] prescribe no such analytical procedures, and only the required precision and trueness of the procedure are defined. With limit value monitoring, the position of the result interval $y \pm U(y)$ is interpreted relative to the limit value (see Figure 3-2).

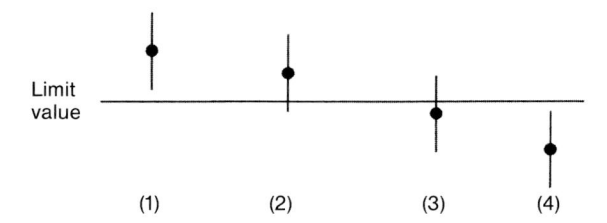

Fig. 3-2 Measurement uncertainty for limit value monitoring.
1) result and measurement uncertainty are above the limit value.
2) result is above, but limit value is within the measurement uncertainty.
3) result is below, but limit value is within the measurement uncertainty.
4) result and measurement uncertainty are below the limit value.

Example (see Table 3-4):

- The limit value is 120 mg/l.
- The measurement uncertainty is $U(y) = 3.5$ mg/l with $P = 95\%$ (commensurate with $\alpha = 5\%$).

Table 3-4 Limit value monitoring.

a) Limit value 120 mg/l is a lower limit value (LV) and result y should not fall below this		b) Limit value 120 mg/l is an upper limit value (LV) and result y should not exceed this	
With the limit value determination the measurement uncertainty 3,5 mg/l is			
unaccounted for	accounted for	unaccounted for	accounted for
y lies above (120 + 3.5) mg/l; limit not reached	y lies above (120 + 3.5) mg/l; limit not reached	y lies under (120 – 3.5) mg/l; limit not exceeded	y lies under (120 – 3.5) mg/l; limit not exceeded
y lies between 120 and 123.5 mg/l; limit not reached	y lies between 120 and 123.5 mg/l with probability of error $\alpha \leq 5\%$; limit not reached	y lies between 116.5 and 120 mg/l; limit not exceeded	y lies between 116.5 and 120 mg/l with probability of error $\alpha \leq 5\%$; limit not exceeded
y lies between 116.5 and 120 mg/l with probability of error $\alpha \leq 5\%$; limit not reached	y lies between 116.5 and 120 mg/l with probability of error $\alpha \leq 5\%$; y falls below limit value	y lies between 120 and 123.5 mg/l with probability of error $\alpha \leq 5\%$; limit not exceeded	y lies between 120 and 123.5 mg/l with probability of error $\alpha \leq 5\%$; y exceeds limit value
y lies under 116.5 mg/l; falls below limit value	y lies under 116.5 mg/l; falls below limit value	y lies above 123.5 mg/l; exceeds limit value	y lies above 123.5 mg/l; exceeds limit value
Conclusion: *falls below limit value* with probability of error $\alpha = 5\%$ at $y \leq 116.5$ mg/l; therefore: $y \leq LV - U(y)$	*falls below limit value* with probability of error $\alpha = 5\%$ at $y \leq 120$ mg/l; therefore: $y \leq LV$	*exceeds limit value* with probability of error $\alpha = 5\%$ at $y \geq 123.5$ mg/l; therefore: $y \geq LV + U(y)$	*exceeds limit value* with probability of error $\alpha = 5\%$ at $y \geq 120$ mg/l; therefore: $y \geq LV$

3.2.7.5 Summary

For the estimation of measurement uncertainty in practice, the following principles apply:

- Refer back to available precision data.
- For a quick, rough estimation, it is sufficient to consider the main uncertainty component(s).
- The measurement uncertainty of the sampling is usually the largest; it can, however, only be roughly estimated.
- The estimation of total uncertainty can also be based on interlaboratory test characteristics.

In general: The measurement uncertainty should not be given as exactly as possible, rather only as exactly as necessary, i.e., it depends on the requirements of the client and the method used.

In every case, the documentation of the procedure characteristics (originally from Phase I) is to be supplemented with details of the measurement uncertainty. Namely, for the result specification consider for all uncertainty components:

- the type of components,
- value of the uncertainty,
- origin and source of this information,
- calculation instructions for the indication of measurement uncertainty.

3.2.8
Reporting Analytical Results

The results of an analysis may only be released to a client if the quality has been proven by means of a plausibility control (see Section 3.5.2.2) and the results of quality controls (see Section 3.3). The final report presents the test results and all other relevant information accurately, clearly, and unambiguously. It should include [128]:

- explicit identification of the report and of all its pages,
- name and address of analyzing laboratory; when different laboratories are involved: their names and addresses and what was done where,
- name and address of the client,
- identification of the sample(s) analyzed,
- date of receipt of sample(s) and dates of analysis,
- identification of analytical procedures used,
- if sampling was carried out by the laboratory: description of sampling procedure,
- any deviations, additions to or exclusions from the standard procedure, and any other information relevant to the procedures used,
- identification of all non-standard procedures utilized,
- presentation of results and their uncertainties, including tables, graphs, etc., and any failures identified,
- identification, including title and signature (with date), of the person responsible for the analysis,
- statement to the effect that the test results relate only to the items tested,
- statement that the complete report may not be reproduced without written consent from the testing laboratory.

3.3
Routine Quality Control

The control chart system developed in Phase II comes into permanent use in routine analysis to assure precision and accuracy. The number and type of the control samples in question are dependent on the process used and the quality targets.

3.3.1
Trueness Control

3.3.1.1 General

A frequently required target in analysis is the accuracy of analytical results. Since accuracy is composed of precision *and* trueness together, the required accuracy cannot be achieved if the result is due to systematic errors, despite attempts at improving precision. Trueness control is therefore an unavoidable aspect of analytical quality control. One possibility for the checking of systematic errors is participation in interlaboratory tests (see Chapter 4). Blank value-, mean value-, and recovery rate charts can serve as internal trueness controls.

3.3.1.2 Blank Value Monitoring

Numerous analytical methods require the determination of blank values in order to correct the measured values of analytical samples by an unspecified amount. A distinction is made between two types of blank samples:

1. Reagent blanks – blanks which contain only distilled water or other relevant solvent and the reagents used in the analytical process in place of the sample.
2. Sample blanks – which contain in place of the sample volume the same volume of a sample of identical matrix but lacking the parameters to be determined.

The use of sample blanks makes sense for the determination of the total blank; however, such sample blanks are not often available in many branches of analysis.

3.3.1.2.1 Causes of Blank Values

Blank values can have many causes, for example:

1. The reagents used are not completely free of parameters to be analyzed.
2. Reactions occur between the matrix components and the reagents which are similar to the analysis reactions (poor process selectivity).
3. The reagents or sample matrix possess a color of their own (which, for example, means that they have their own absorptions in photometric evaluation).
4. Systematic contamination of reaction vessels and measuring instruments.
5. Differences between cuvettes in photometric measurements.
6. Decomposition of the reagents, for example through aging or incorrect storage conditions, etc.

3.3.1.2.2 Practical Use of Blank Samples and Blank Values

It is possible to evaluate measured values in the laboratory by various means:

The measured value of the analytical sample is corrected by the amount of the blank sample (blank subtraction). The concentration is then determined using a corrected calibration function [161].

1. The reliability of such an evaluation must be questioned, since an intercept, i.e. a blank value a, is calculated by the determination of the calibration function. If the calibration curve is then corrected for this intercept, i.e. the calibration curve begins at $a = 0$, information about equipment blanks, reagents, etc., is lost.
2. In the case of photometric methods, the sample is often measured directly against the blank. By this still frequently used procedure, information about the quantity in question and possible fluctuations of the blank value is lost. This can lead to a considerable systematic, but undetectable, falsification of the analysis results [29].
3. The *recommended procedure* for the determination of analytical results is evaluation with an uncorrected calibration curve (see Section 1.2.3). The blank value measurements in question do *not* show up in the result, but instead serve only as a basis for quality assurance measures.

For example, in photometric analysis, the measured values for both the blank and the analysis sample are determined by measuring against distilled water or other pure solvent. The measured blank values are then entered on a blank value control chart.

3.3.1.2.3 Blank Value Control Charts

The blank value control chart (see Section 2.6.7.1.5.2) can detect

- impurities in the reagents,
- impurities from reaction vessels and measuring instruments,
- instrument errors (e.g., baseline drift).

Therefore, it is prudent to carry out a blank value determination at the beginning and end of each analysis series. The blank values obtained are entered on the blank value control chart.

The establishment of decision-making criteria for the blank value control is difficult because a Gaussian (or normal) distribution for blank values may only be assumed to a limited degree.

A normal distribution of blank values would mean that negative results are possible, but these are, as a rule, not expected by measurement. Until recently, the literature has been lacking in more exact information about the distribution and control of blank values.

It is therefore recommended to establish a blank value control chart using the same decision-making criteria as for a mean value chart for use as a temporary blank value control. After a sufficient number of blank value determinations, the blank value distribution can be calculated and new decision criteria can be established.

The use of mean value chart decision criteria appears to be provisionally justified since Inhorn [117] and Burr [14] state that $\bar{x}$-charts, which require a Gaussian (normal) distribution of values, may also be more broadly used for data that are not normally distributed. In addition, Burr lists a series of factors that can be used to determine the control and warning limits for non-normally distributed values; of course, this arrangement assumes knowledge of the respective distribution form.

3.3.1.3 $\bar{x}$-Chart

Standard solutions, synthetic samples, or certified reference materials may be used for accuracy control by means of $\bar{x}$-charts (see Section 2.6.7.1.5.1). Of course, quality assurance by way of standard solutions serves only for the verification of the calibration and/or the measurement instrument.

If, on the other hand, samples of synthetic or actual matrix are used for trueness control, then the selectivity of the analytical process in use may also be verified.

The control sample in question should be analyzed at least once per analysis series and the results entered into an $\bar{x}$-chart.

The analysis of a certified reference material is preferred for accuracy control. However, the availability of certified reference materials is very limited in many branches of analysis and the diversity of actual matrices is very large. In addition, these samples are usually very expensive. Therefore, they are co-analyzed only rarely. Instead, a limited accuracy control is often performed using a recovery rate control chart.

3.3.1.4 Recovery Rate Control Chart

The recovery rate control chart (see Section 2.6.7.1.5.3) certifies the trueness of analytical results while simultaneously including matrix influences.

Since very different matrix types are analyzed (for example, in water analysis there is surface water, communal waste water, and industrial waste water), it is necessary to maintain a separate control chart for each type of matrix.

A recovery rate control chart provides only limited trueness control since only matrix-dependent proportional systematic deviations and not constant systematic deviations can be detected during the determination of the recovery rate.

A meaningful result can only be obtained if the original sample is spiked with a suitable substance. For example, maintaining a recovery rate control chart for COD in water analysis makes little sense if potassium hydrogen phthalate is added to the original sample since this substance is easily oxidized and one therefore ends up verifying only the correct dosage of spiking solution [158].

3.3.2
Precision Control

3.3.2.1 General

In routine analysis, the precision of results can be checked in four different ways:

1. Using an R-chart (see Section 2.6.7.2).
2. Using a standard deviation control chart (see Section 2.6.7.4).
3. By multiple determinations.
4. By the method of standard addition (see Section 3.4.1.1).

3.3.2.2 Precision Control Using an *R*-Chart

The precision of analytical results in routine analysis can be determined with very little additional analysis effort by means of *R*-charts (see Section 2.6.7.2).

For this, duplicate determinations are performed in each analysis series for an arbitrary but typical actual sample. For samples with very different matrices and large concentration differences, the maintenance of several *R*-charts for different matrix types and concentration regions is recommended. The standard deviation for a certain analytical result can be estimated from the *R*-chart, provided that the matrix of the sample in question is the same as that of the control sample chosen for the *R*-chart. As a safeguard, the range of the sample in question should be determined and entered in the control chart in order to prove that an out-of-control situation does not exist.

The standard deviation is estimated from the mean range of the *R*-chart as follows:

$$s = \frac{\bar{R}}{d_2} \text{ with } d_2 = \text{factor, see Table 3.5}$$

d_2 is dependent on the number of measurements, n, within a sample; $n = 2$, or duplicate determinations, is most common for chemical analyses.

Table 3-5 Factors for calculating the standard deviation from the mean range [4].

Size of subgroup	Factor d_2
2	1.128
3	1.693
4	2.059
5	2.326
6	2.534
7	2.704
8	2.847
9	2.970
10	3.078

The distribution determined in this way can be used to evaluate whether an obtained result possibly exceeds a set limit.

3.3.2.3 Securing Precision Using a Standard Deviation Control Chart

A standard deviation control chart (see Section 2.6.7.4) may be used in the same way as an *R*-chart for precision control.

3.3.3
Revision of Quality Control Charts [136, 196]

Control charts are checked over certain time intervals for changes to the mean values and the warning and control limits. A revision of the last 60 data points is carried out. If during this revision there are one to six cases in which the $2s$ warning limits were exceeded, then there is probably no change in the precision. A redefinition of the boundaries is not necessary.

However, if either none or more than six control values are situated outside of the warning limits, then with 90% certainty the precision has changed (improved or deteriorated), that is to say, revision of warning and control limits is necessary. In this event, it must be decided whether or not the central line and the warning and control limits of the control chart should be recalculated on the basis of the last 60 data points. If new, broader control limits are to be calculated, attention must be paid to relevant precision demands for the given substance.

3.3.4
Quality Assurance in the Case of Time-Consuming or Infrequent Analyses

Time-consuming analytical procedures usually result in very small analytical batches, sometimes consisting of only one analysis. In order to avoid quality problems resulting in additional effort for repetition of analyses, quality assurance measures in the preparatory phase and during routine analysis become extremely important. The calibration characteristics must be valid; recovery rates and precision must be proven. Economic considerations do not justify any lapses in quality control analysis.

Analyses that are not performed regularly, for example only once or twice per week or per month or even just a few times during the year, must be accompanied by:

- a check of the current calibration function,
- a check of the precision by repetition of the analysis and the maintenance of a range control chart,
- a check of the recovery rate by the standard addition method and the maintenance of an RR-control chart,
- an analysis of a blank sample and the maintenance of a blank value control chart,
- an analysis of a reference material to check the recovery rate and the maintenance of an RR-control chart.

Compared to large daily batches of identical analyses, extremely small and/or infrequent batches are subject to several quality problems that must be overcome by appropriate measures:

- the staff has limited experience; in the case of seldom applied analytical procedures, even fluctuation of staff and "no experience" must be considered,

- invalidity of the previous calibration,
- perishability of reagents,
- perishability of control samples,
- unobserved deadjustment of equipment.

As quality assurance measures may comprise more than 50% of the total analytical cost, one should consider subcontracting out infrequently used analytical methods and analyses to a specialized laboratory.

3.4
Special Quality Problems in Routine Analysis

3.4.1
Matrix Effects

Environmental analysis deals with great variation in the composition (matrix) of samples; for example, there is no "standard" matrix for communal waste water. Therefore, it is problematic to make a prognosis for the imprecision (s) of a future measured value from the estimated standard deviation of an analysis series investigated at an earlier time with an arbitrarily chosen matrix.

Even more problematic is the calculation of the prognosis interval using a previously determined calibration function. This is especially true if the entire process (e.g., including digestion, extraction, etc.) was not calibrated and standards in pure solvents were used as calibration standards for the parameters to be determined; measured values of calibration samples and actual samples do not possess the same statistical structure.

Therefore, the imprecision of an analytical result should be determined or estimated using the method of standard addition, especially if monitoring a limit. (Of course, the trueness of the analytical work must also be monitored; see Section 3.3.1.)

3.4.1.1 Determination of Precision Using the Method of Standard Addition
Estimation of the standard deviation using a range chart or multiple analyses records a matrix-dependent imprecision but not a possible untrueness, i.e. a matrix-dependent systematic error.

If the method of standard addition [33] is applied, not only can the precision be determined but the trueness of the result may also be improved.

3.4.1.1.1 Method of Standard Addition
If possible, the method of standard addition should only be applied to standard analytical processes for which the region of linearity is known, the variance homogeneity over the range is proven, and a calibration function exists [99, 112, 199]. For the spiking, the original sample is divided into at least N partial samples of equal volume. $N-1$ of these samples are then spiked.

Note: It is advisable to make each sample up to a constant volume with the solvent (e.g., distilled water) after each spiking step. The unspiked original sample should also be diluted to this volume.

a) Determination of Spiking Concentrations

The sample should be spiked so that the analyte concentration is doubled. The measured value y_0 of the bulk (or gross) sample is determined first and the sample content x_0 is estimated by means of the known calibration function. The bulk sample is then spiked in four equidistant concentration steps so that the maximum spiking concentration x_{s4} is the same as x_0 [178]. The concentration of the most-spiked sample $(x_0 + x_{s4})$ should lie within the working range of the linear calibration function.

Note: Routine use of four spiking concentrations represents a compromise between the desired accuracy of the analytical results and acceptable analysis effort.

b) Preparation of Standard Solutions

The standard solution should be constituted so that spiking can be carried out in steps of equal volume, i.e. the first spiking volume V_s increases the concentration by x_{s1}, the second spiking volume $2 \cdot V_s$ increases the concentration by $x_{s2} = 2 \cdot x_{s1}$, and so on.

In order to avoid matrix changes, the standard solution containing the analyte should be prepared so that the spiking volume remains as small as possible in relation to the sample volume.

c) Determination of Measurement Values and Evaluation

All samples (original samples and spiked samples) are prepared according to the analytical protocol (e.g., including digestion). The measured values y_0, y_{s1} to y_{s4} are determined for these samples. A blank value measurement is performed as well.

A spiking calibration function with appropriate confidence interval is calculated from the data pairs (without including the measured blank value) through simple linear regression. The spiking calibration function is extrapolated to the abscissa. The blank value y_B is needed for the calculation of the sought concentration $\hat{x}$ (see Figure 3-3).

Spiking calibration function:

$$y = \hat{y}_0 + b_S \cdot x \tag{118}$$

$\hat{y}_0 =$ axis intercept of the spiking calibration function

Sample concentration:

$$\hat{x} = \frac{\hat{y}_0 - y_B}{b_S} \tag{119}$$

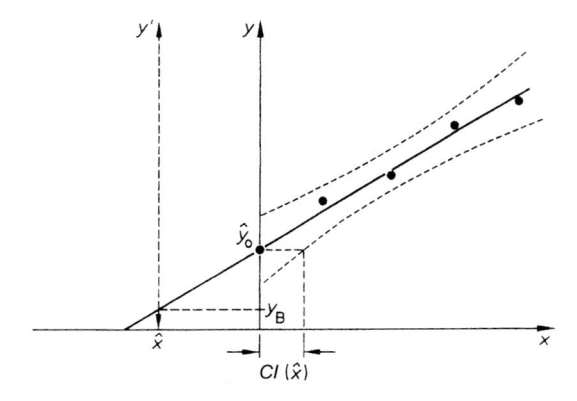

Fig. 3-3 Estimation of the sample content ($\hat{x}$) with the appropriate confidence interval for a spiking experiment.

If during the sample preparation "filling up" of the volume is required, the appropriate corrective calculation (see Section 2.5.2.4) must be applied as follows:

$$\hat{x} = \frac{\hat{y}_0 - y_B}{b_S} \cdot \frac{V_{sample} + V_{added}}{V_{sample}}$$

The confidence interval is determined at the point $\hat{y}_0$ (see Figure 3-3):

$$CI(\hat{x}) = \frac{s_y}{b_S} \cdot t_{f,P} \cdot \sqrt{\frac{1}{N} + \frac{1}{N_a} + \frac{(\hat{y}_0 - \bar{y})^2}{b_S^2 \sum (x_i - \bar{x})^2}} \quad \text{where } f = N - 2, P = 95\% \text{ and } N_a = 1 \quad (120)$$

If the confidence interval of the spiking calibration function has to be narrowed, then multiple-N_a-determinations may be made at each spiking step. Each individual spiked sample must undergo the entire analytical process. It is also possible to increase the number of standard additions.

3.4.1.1.2 Statistically Securing Analytical Results

Since the matrix influence of a sample not only reveals itself as a systematic error, but may also considerably worsen the precision of the results, it must be determined whether the obtained result differs from the concentration $\hat{x} = 0$.

To do this, the test value x_a is calculated (see Section 1.2.4.2.3) [196].

$$x_a = 2 \cdot \frac{s_y}{b_S} \cdot t_{f,P} \sqrt{\frac{1}{N} + \frac{1}{N_a} + \frac{(y_a - \bar{y})^2}{b_S^2 \sum (x_i - \bar{x})^2}} \quad (121)$$

where

$$y_a = \hat{y}_0 + s_y \cdot t_{f,P} \cdot \sqrt{\frac{1}{N} + \frac{1}{N_a} + \frac{\bar{x}^2}{\sum (x_i - \bar{x})^2}} \quad (122)$$

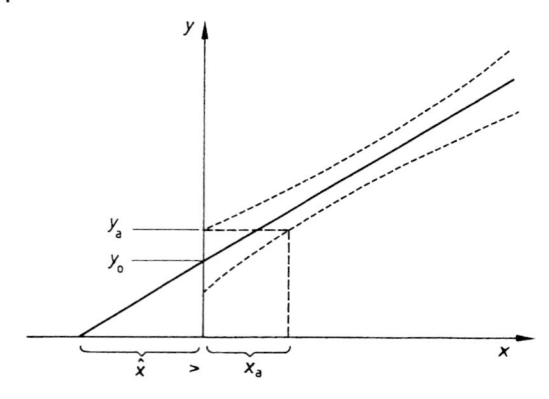

Fig. 3-4 Determination of the test value x_a.

Decision: For $x_a < \hat{x}$ (see Figure 3-4), the calculated concentration $\hat{x}$ is significantly different from the concentration $x = 0$.

For $x_a > \hat{x}$, the calculated concentration cannot be significantly differentiated from the concentration $x = 0$ as a result of too large an imprecision.

3.5
Corrective Measures

3.5.1
Sources of Error in Analytical Laboratories

Every analytical result contains unavoidable error. Knowledge of and attention to possible sources of error help to considerably reduce the occurrence of error and also to locate more rapidly the causes of an error that has appeared (troubleshooting).

Error prevention is primarily achieved by optimizing the operational process. The better a laboratory is organized, the fewer errors can be expected. Optimal laboratory organization and quality assurance corroborate each other reciprocally [109].

If an analytical process is used routinely, then errors that appear may be categorized as follows:

1. sample errors
2. reagent errors
3. reference material errors
4. method errors
5. calibration errors
6. equipment errors
7. signal registration and recording errors
8. calculation errors
9. transmission errors
10. errors in the interpretation of results

One must also differentiate between three types of errors [109, 191]:

Gross Errors
A very large (gross) deviation appears in the results due to the presence of a gross error. Gross errors can appear randomly as so-called "outliers". Primary causes of these are confusion, random calculation errors, errors in transmission of information, misspellings, etc. Repeated incorrect preparation of standards and samples, incorrectly chosen analytical processes, systematic deviations from prescribed procedures (modification of analysis steps, use of incorrect equipment, containers, etc.), defective or incorrectly adjusted equipment, systematic arithmetical errors, and so on, lead to gross errors for all results of one or more analysis series and can be detected by control analyses using an $\bar{x}$-chart. It is difficult to distinguish between the gross errors mentioned above and systematic errors.

Systematic Errors
The analytical results show systematically high or low values. The accuracy of the results decreases.

Random Errors
The precision of the results decreases, the scatter increases.

A faulty result cannot always be categorized strictly as being due to one type of error. A gross error is often only distinguishable from a systematic error by the extent of its influence on the result, or it consists of a combination of considerable systematic and random errors.

Table 3-6 indicates types of error and possible causes.

3.5.2
Systematic Troubleshooting

The recognition and elimination of gross and systematic errors is the primary task of routine quality assurance. The recognition of errors includes three steps:

1. Checking results for plausibility as a matter of course (see Section 3.5.2.2).
2. Quality monitoring using control charts (see Section 2.6.7).
3. Inspecting (and maintaining) the workplace and equipment to uncover errors before they are apparent in the analytical results.

To a large extent, successful troubleshooting depends on experience and knowledge of the system. Therefore, sources of error and their elimination should be documented and discussed.

Table 3-6. Examples of common sources of error in an analytical laboratory; aids for troubleshooting [12, 13, 16, 108, 191].

		Cause of error	Type of error	
			systematic	random
1		**Sample errors**		
1.1		*Sampling errors*		
		• contamination of the sample (e.g., by the container, Cu and Zn from brass pipes when taking water samples)	×	
		• loss	×	
		• incorrect sample container (e.g., O_2 diffusion from sampling lines)	×	
		• non-representative sample (e.g., unfavorable sampling point, insufficient mixing (e.g., sewage sludge fermentation tank), poor timing for the preparation of an averaging sample)	×	
		• incorrect sampling place (e.g., wrong patient in medical diagnosis, wrong body of water in water analysis, sample taken at incorrect place)	×	
1.2		*Errors in transportation and storage of samples*		
		• evaporation (sample container is not properly closed)	×	
		• contamination from the outside (e.g., for organic vapor in air, heavy metals in laboratory dust)	×	
		• storage at an incorrect temperature (e.g., decomposition of organic substances)	×	
		• disregarding of the light sensitivity of some substances (e.g., PAHs)	×	
		• disregarding of the instability (physical, chemical, biochemical) of the components	×	
		• incorrect preservation of or failure to preserve samples	×	
		• unacceptable aging of the sample (analysis is performed after a long weekend or even after a vacation period)	×	
		• incorrect storage container (insufficiently cleaned container, uptake of substances from the container; e.g., softeners in plastic containers, crumbling rubber stoppers, etc.)	×	
1.3		*Errors in sample identification*		
		• switching		×
		• incorrect time or quantity information (e.g., in the collection period)	×	×
1.4		*Errors in sample preparation*		
		• blank values (e.g., from grinding equipment material for sewage sludge samples)	×	
		• dilution errors	×	×
		• volumetry or balance errors	×	
		• inhomogeneity, distribution errors (e.g., incorrect timing for the separation of solids, insufficient homogenization before or during the measuring or dosing)	×	

Table 3-6. (continued)

	Cause of error	Type of error	
		systematic	random
	• losses (e.g., evaporation of volatile substances)	×	
	• sample digestive conditions that cannot be reproduced (e.g., inaccurate time or temperature maintenance during the digestion)		×
	• sample contamination (see errors in sampling (1.1) and errors in transportation (1.2))	×	
2	**Reagent errors**		
	• impure reagents	×	
	• impure solvents	×	
	• improper storage of reagents	×	
	• use of older or out-of-date reagents (e.g., neglect of expiry date)	×	
	• incorrect solvent volume	×	
	• reagents are incompletely dissolved	×	
	• evaporated reagents	×	
3	**Reference material errors** (control samples, standards)		
	• impurity of reference materials	×	
	• errors from interfering substances in the reference material	×	
	• physical differences between the sample and the reference materials (e.g., viscosity)	×	×
	• incorrect reference value used	×	
	• reference values are inexact (e.g., inexact means, standard deviations from too short a preliminary period)		×
	• changes in the reference materials during (inappropriate) storage (concentration decreases, e.g., through adsorption or decomposition; concentration increases, e.g., through evaporation)	×	
	• use of expired reference materials	×	
	• errors in the preparation of reference materials (e.g., pipetting errors)	×	×
	• insufficient mixing of thawed reference materials	×	×
4	**General method errors**		
	• deviation from the analysis protocol	×	
	• reaction times are not observed exactly	×	×
	• arithmetic errors in the preparation of mixtures, dilutions, additions	×	
	• use of an incorrect analytical process (unsuitable process, a process unacceptable for the analytical objective based on its imprecision; e.g., use of an orientation test instead of the required reference procedure)	×	

Table 3-6. (continued)

Cause of error	Type of error systematic	random
• exceeding or not attaining the validated range	×	
• disregard of the linear range	×	
• disregard of the decision limit or the limit of quantification	×	
• disregard of a necessary sample or reagent blank value	×	×
5 Calibration errors		
5.1 *Inexact composition of calibration standards*	×	
5.2 *Volumetric measuring errors (see 6. Equipment errors)*	×	×
5.3 *Weighing errors*		
• jolting of the balance		×
• inexact zeroing	×	
• soiling of the balance container	×	
5.4 *Inaccurate equipment adjustment*	×	
6 Equipment errors		
6.1 *General errors*		
• equipment used (including glassware and disposable articles) was not sufficiently clean:		
– contamination	×	×
– insufficiently dried equipment	×	×
• use of unvalidated equipment; equipment not recalibrated; adjustment forgotten	×	
• maintenance neglected	×	
• physical influences on instruments:		
– external temperature	×	×
– external electrical and magnetic fields	×	×
6.2 *Errors from the use of glass pipettes*		
• switching of pipettes (incorrect volume)	×	
• use of unsuitable pipettes (e.g., too large a pipette for the volume)	×	×
• use of wet pipettes	×	
• damaged pipette tips	×	
• use of pipettes that are not or cannot be standardized	×	
• incorrect pipetting technique: (e.g., disregard of the required draining time, inexact adjustment of the meniscus, disregard of „IN, EX" markings, pipette was not held vertically, air bubbles were drawn up)	×	×
• insufficient surface wetting as a result of insufficient cleaning	×	×
• hasty work		×

Table 3-6. (continued)

Cause of error	Type of error	
	systematic	**random**
6.3 *Errors from the use of automatic volumetric pipettes*		
• switching of pipettes (incorrect volume)	×	
• volume errors (control by weighing)	×	×
• pipette tip incorrectly attached	×	×
• soiling of the bulb shaft: (e. g., pipette hub jerked backwards, fluid entering the barrel shaft, pipette stored lying down with a wet tip)	×	
• use of leaky pipettes or tips	×	×
• incorrect and hasty work (e. g., pipette hub jerked backwards; tip is not attached in an airtight manner; carryover errors, new tip for every solution; fluid attached on the outside; too rapid ejection of fluid; air bubbles were drawn up)	×	×
6.4 *Dosing with automatic diluters or dispensers*		
• sample needle is clogged		×
• use of incorrectly calibrated instruments	×	
• leaky valves or tubes	×	×
• clogged valves or tubes	×	×
• dosage volumes incorrectly adjusted	×	
• hasty work		×
• insufficient mixing, homogenization		×
6.5 *Inaccurate temperature regulation*		
• incorrect temperature program	×	
• incorrect target temperature	×	
• necessary waiting period not adhered to	×	
• too much or too little fluid in the thermostat	×	×
• inhomogeneous temperature (e. g., caused by defective agitator, circulation pump)	×	×
• with flow thermostat: pump defective or not switched on, valve shut	×	×
6.6 *Cuvette errors*		
• cuvette defects (varying transparency) are not considered, different cuvettes are used interchangeably without zeroing in between	×	×
• use of incorrect cuvettes (e. g., incorrect thickness, unsuitable cuvette glass)	×	
• cuvettes inserted incorrectly	×	
• minimum fill volume disregarded	×	×
• use of cuvettes which are wet outside	×	×
• use of soiled or scratched cuvettes	×	×
• air bubbles or streaks in the solution to be measured		×
• carryover effects	×	
• improper cleaning (e. g., for non-aqueous solutions)	×	

Table 3-6. (continued)

Cause of error	Type of error	
	systematic	random
6.7 *Photometer errors*		
• errors in the adjustment of wavelength	×	
• insufficient light intensity (e.g., outdated lamps)	×	
• dirty optical system (mirrors and lenses which are dusty or clouded by vapors)	×	
• use of improper photometer (filter photometer instead of spectral or spectral line photometer, and therefore – a lack of monochromacy	×	
• photometer is incorrectly or not at all adjusted	×	
• drift effect is disregarded	×	
• incorrectly set zero point	×	
• light entering the sample chamber	×	×
• incorrect program or factor for digital photometers	×	
• incorrect shutter opening (scattered light)	×	
• wrong filter or wavelength setting	×	
• lamp has not stabilized after turning on	×	
• wrong light source in use	×	
• photomultiplier is damaged or out of adjustment	×	
• light source has worn out or failed completely	×	
• unfavorable reading range		×
• noisy recorder		×
• unstable reading due to incorrect electrical installation		×
• interference through electromagnetic pollution		×
6.8 *Transportation or pump systems* (e.g., for flow systems, AAS instruments, flame photometers, auto-analyzers)		
• incorrect program chosen	×	
• clogged tubes	×	×
• leaky tubes or links	×	×
• contaminated tubes	×	×
• errors with the media (e.g., incorrect gas, incorrect medium quality, contamination, i.e., through additives during maintenance)	×	
6.9 *Errors in the atomizer* (e.g., for AAS instruments, flame photometers)		
• clogged	×	×
• mechanically defective	×	×
• incorrectly adjusted	×	×
6.10 *Errors in the spectroscopy*		
• disregard of line overlap in emission spectroscopy	×	
• spectrometer not calibrated	×	
• incorrect evaluation program	×	

Table 3-6. (continued)

	Cause of error	Type of error	
		systematic	random
7	**Signal registration and recording errors**		
	• choice of incorrect range	×	×
	• choice of incorrect measurements	×	
	• reading errors		×
	• switching of data		×
	• recording errors		×
8	**Calculation errors**		
	• arithmetic errors, decimal point errors, incorrect units		×
	• rounding errors		×
	• disregard of the blank value of reagents or samples	×	
	• use of an inexact, unverified absorption coefficient	×	
	• neglect of or use of an incorrect dilution factor	×	
	• inattention to a switch-over in the measured range		×
9	**Transmission errors**		
	• assignment errors, mix-ups		×
	• misspellings		×
	• transfer errors (transmission of distorted or incomplete results)		×
10	**Errors in the reporting of results**		
	• omitting a sample error		×
	• assignment of a result to the wrong normal range		×
	• misinterpretation of units and magnitudes		×
	• doing without quality assurance measures	×	×
	• ignoring „out-of-control" situations	×	×

3.5.2.1 Analytical Errors That Can Be Detected Using Statistical Quality Control Methods

If an out-of-control situation appears on a control chart, then it must be concluded that the same error exists for the analysis of actual samples. A quick recognition and elimination of the existing error is absolutely necessary for the certainty of the analytical results.

Procedures for quick and successful troubleshooting in analytical chemistry are suggested in the following sections [191].

3.5.2.1.1 Elimination of Gross Errors

The control sample analysis is repeated while avoiding possible gross errors (see Table 3-6) and strictly following established procedures. If the control sample results show that the process is again in control, then presumably the analysis protocol was not followed exactly for the previous analysis series or a gross error existed. The entire series of analyses must be repeated after elimination of this error.

On the other hand, if the analytical results of the control sample are erroneous but reproducible, then a systematic error probably exists.

3.5.2.1.2 Elimination of Systematic Errors

To test for systematic errors, several different accuracy control samples are analyzed, if possible. In order to recognize reagent- and method-dependent errors, the concentrations of these control samples should be distributed over the entire measurement range. As a minimum, an accuracy control sample should be used at the lower and upper portions of the range. These additional control sample analyses can be omitted if the control charts used already indicate systematic errors (trends). If a systematic error exists, i.e., if the results of the control sample lie predominantly above or below the expected values, then one should look for possible causes step-by-step using a source of error table (see Table 3-6). Exchange experiments are recommended wherever possible; many errors can be detected quickly by exchanging individual reagents, equipment, or personnel.

3.5.2.1.3 Improving Precision

To improve precision, one can also go through Table 3-6 step-by-step in order to determine possible causes of higher imprecision.

Often, the dosage equipment and the working methods of the operators are causes of poor precision. In the worst case, the (im)precisions of the individual steps of a process add up according to the law of error propagation and are then known as the total (im)precision [109]. If the total precision of a method is to be improved, then those analytical steps which have the greatest influence on the total error of an analytical process must be identified first.

Figure 3-5 represents a sample troubleshooting scheme. Troubleshooting schemes should if possible be created by the manufacturers of analytical instruments and the originators of analytical processes and be provided to users since it is impossible to create *one* generally applicable troubleshooting scheme for *all* analytical processes. A valuable aid is the procedure described in Phase I for the stepwise testing of an analytical process.

Errors may also appear that cannot be detected using statistical quality assurance methods. These are mostly errors that only affect one analysis and do not influence the next sample (e.g., gross, matrix- or sample-dependent errors, errors of transmission, and outcome of results). These errors can only be discovered using plausibility checks.

3.5.2.2 Plausibility Checks

"Plausibility is a measure of how close a sample value comes to the expected value" [158]. An effective plausibility check can only be conducted if suitable pre-

3.5 Corrective Measures

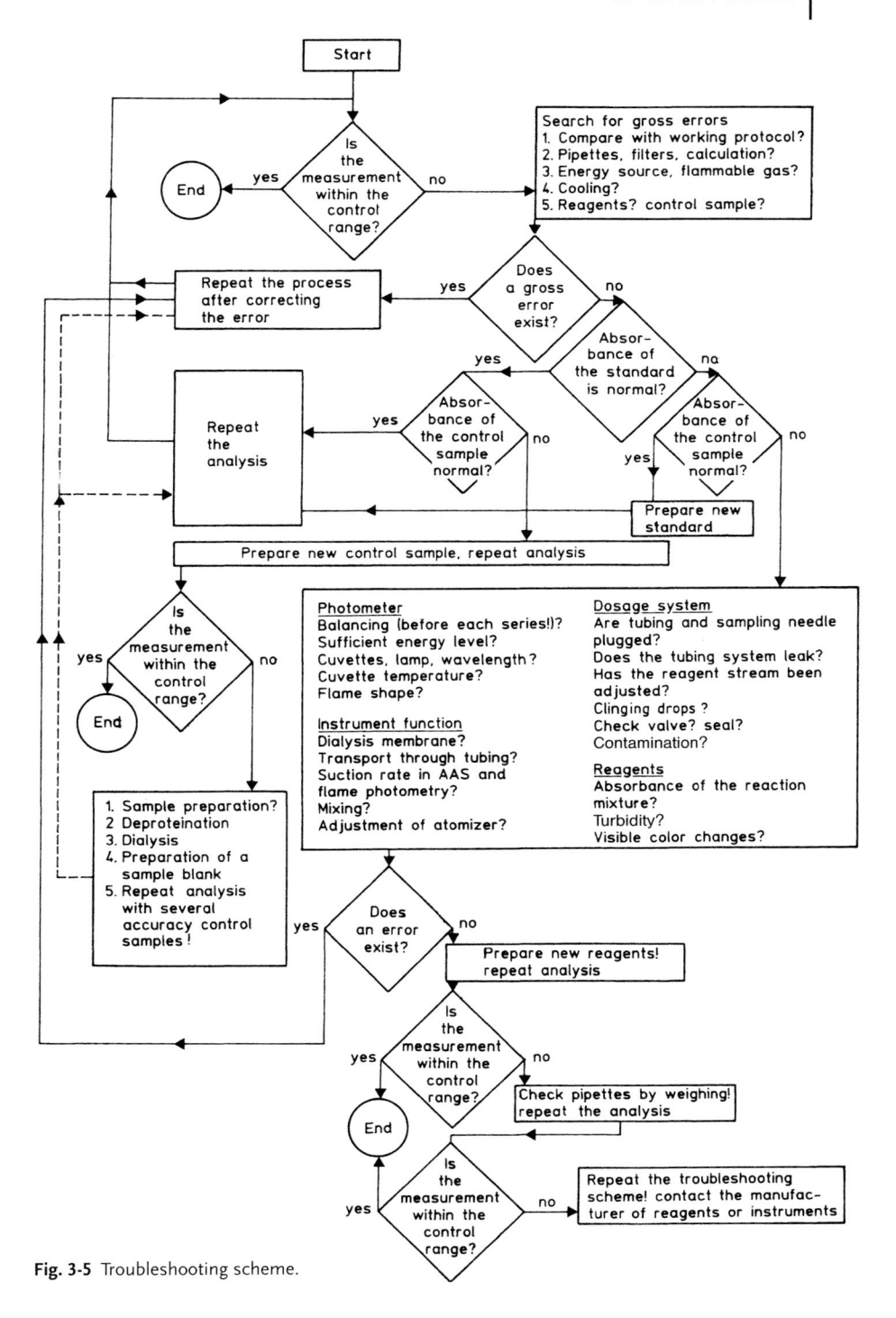

Fig. 3-5 Troubleshooting scheme.

conditions and/or information is made available. The plausibility check process is composed of two parts:

1. information/input,
2. check, see chart (Figure 3-6).

3.5.2.2.1 Information/Input

Specific background information to be provided by the client includes:

- Information about the reason for the investigation, e.g., routine sampling according to schedule XY, incident, special problem, etc.

- Information about the origin of the sample, for example:
  a) For waste water samples:
     The type of water treatment plant the sample comes from, its production processes, the name of the company, and if necessary technically competent contact persons (identification by code for confidentiality reasons), type and quantity of raw materials, products, stored substances, etc., that are relevant to the investigation.
  b) For surface water samples:
     Description of the body of water and the place from which the samples were taken, notes about possible influences from direct sources or seepage from old deposits or landfills.
  c) Other samples, analogous to a) and b).

Furthermore, the investigating laboratory must also be informed of known limit requirements or analytical quality targets so that the method and investigative effort can be matched to the problem in question.

If the client did the sampling, then it is absolutely necessary that the following sample data are given to the laboratory:

- place of measurement, place the samples were taken,
- for waste water: type of source (directly from production or pre-cleaned or diluted, indirect),
- type of sampling (random, mixed sample, sample proportional to waste stream, etc.),
- anything remarkable about the sample,
- on-site preservation, sample pre-treatments (homogenization, sedimentation, etc.),
- results of on-site measurements (pH, conductivity, oxygen, temperature, etc.)
- field test results.

Based on experience, the common practice of just assigning numbers to samples and sending them along with a list of parameters to be investigated to a laboratory can lead to the use of unsuitable analytical processes and misleading results, since the analyst does not know the question being asked and is not in the position to perform a plausibility check based on background information and characteristic sample data.

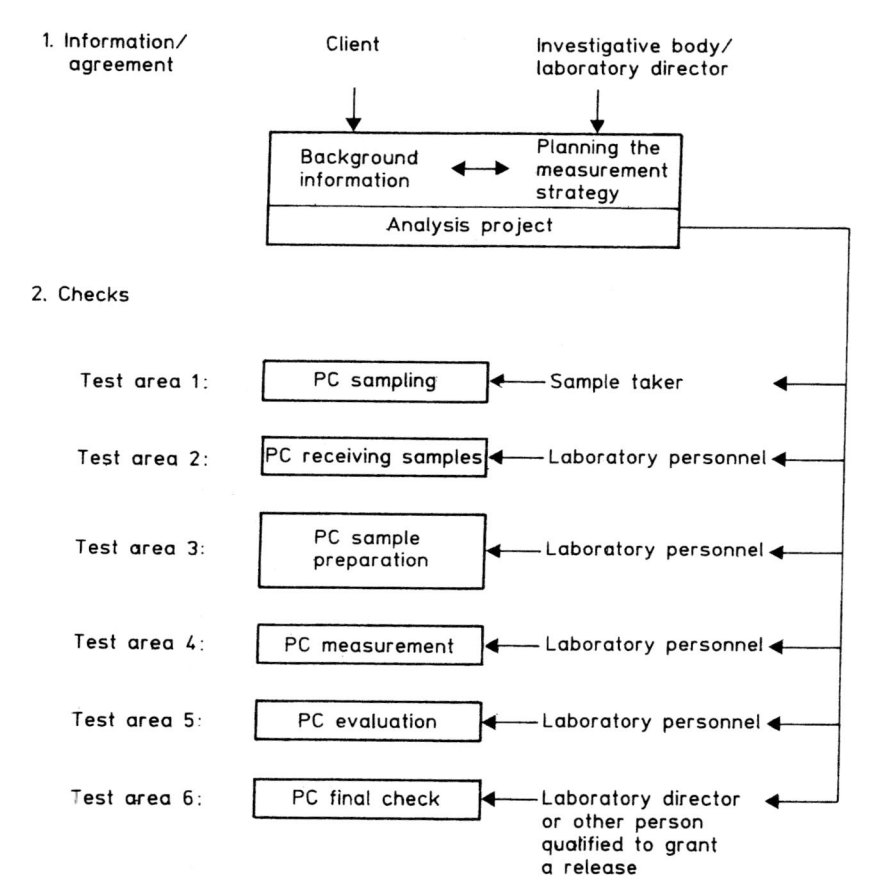

Fig. 3-6 Flow chart for a plausibility check (PC) [158]. The project being analyzed is the entire analytical process including measurement planning and evaluation of results.

3.5.2.2.2 Check

The plausibility check for test areas 1 to 3 in Figure 3-6 is not within the scope of this book.

Methodical and statistical verifications are preferred for test area 4 (measurement) (see Section 3.5.2.1).

In the analytical laboratory, the actual plausibility check for the evaluation of analytical results occurs in test area 5. Here, the verification of atomic and/or ion balances, in water analysis for example, plays a decisive role:

- nitrogen balance,
- phosphorus balance,
- carbon balance,
- total ion balance.

The interrelationship of various parameters must also be examined, for example:

- COD/BOD_5 relationship,
- Cr, total/Cr(VI),
- Fe, total/Fe(II),
- electrical conductivity/sum of ions,
- dissolved organic carbon (DOC)/chemical oxygen demand (COD),
- adsorbable organic halogens (AOX)/sum of halogenated hydrocarbons.

Computers may be used for these plausibility investigations.

In test area 6 (final control), the analysis results are verified based on the background information provided as well as the sample data. This type of plausibility check requires access to extensive data. The management of computer-supported client databases is recommended.

The following points, for example, need to be considered:

- Are the type and composition of waste-water ingredients typical of the process being monitored (e.g., water, landfill)?
- Do the analytical data "fit" earlier findings?
- In clinical analyses: does the pattern of results from one patient correspond with (patho)physiology, biochemistry, etc.?
- Where do the expected values lie?

The cause of an implausible measurement result must be determined (perhaps using a scheme) and those responsible for analytical quality assurance consulted.

Furthermore, the statistical data must be verified (process standard deviation, control chart, etc.).

3.5.2.2.3 Measures and Consequences Arising from the Plausibility Check

If the error cannot be determined and corrected, then the analysis must be repeated. If possible, a second independent analysis process should be used. If the results are confirmed, then the client must be notified. The cause should be discussed together with the client (e.g., operational incident).

3.6
Documentation and Archiving

The documentation is not an end in itself, but enables, within the framework of quality management [66], intentions and the consistency of measures to be proven in a possible later disagreement. Documentation and archiving time periods can also be legally prescribed [9, 10]. According to the quality manual, the following information must be recorded and archived:

- plans of the laboratory with regard to construction, installation, environmental protection, fire and civil protection and equipment,

- all operating instructions,
- all SOPs and procedure descriptions,
- documentation of the waste disposal, in particular of special waste,
- documentation of personnel training activities and briefings,
- documentation of all quality control and quality assurance measures and their results,
- specification of the materials used, quality assurance verification from the suppliers,
- sampling plans and reports,
- book-keeping of the stored materials and samples,
- raw data, measured values, and analytical results; analytical reports,
- laboratory reports,
- final project reports,
- reports of the results of the interlaboratory tests in which the laboratory has taken part,
- external audit reports,
- quality assurance reports: internal inspections and checks, their results and actions/measures,
- instrument log books with reports of all inspections, calibrations, preventive maintenance, and repair activities,
- specifications and operation manuals of instruments and software.

In addition, general conditions are to be included in all documentation of an action or measure:

- in what context and for what purpose were the measures performed, i.e., who commissioned them,
- when the measures were performed,
- who performed the measures,
- what were the measures carried out with; this includes information within the framework of calibration of measuring equipment allowing tracing of the performance of measuring instruments to the national standards,
- extraordinary circumstances that could influence the measures or the results.

As long as they are stable, any leftover analytical samples, and calibration, quality control, and reference materials should be retained.

4
Phase IV: External Analytical Quality Assurance

4.1
Introduction

If the routine quality assurance measures described in Chapter 3 are performed by analytical laboratories, then it is to be expected that these laboratories perform the analytical process in question *reliably* and that results from these laboratories should be *comparable*.

Different external measures used to check and prove a laboratory's quality system have different objectives:

- Audits [64, 66] are systematic and independent evaluations by external auditors who check if quality-related activities and their results comply with the described procedures and whether these procedures efficiently lead to the quality objectives. The quality objectives themselves are not evaluated.
- Analytical assessments in the form of interlaboratory trials are carried out with two objectives: the assessment of the analytical procedure during the standardization process and the evaluation of the analytical performance of the laboratory in routine analysis and/or as part of an accreditation procedure [127].

4.2
Audits [66, 122]

Audits may be performed internally or externally. Internal audits are part of the quality assurance system. They are carried out regularly by a competent independent member of the staff (in areas where he or she is not personally involved) or by an external consultant. The aim of an internal audit is the assessment of the current quality assurance system and the proposal of necessary corrective measures and alterations. Besides scheduled internal audits, additional partial or total quality audits should be performed in cases of severe quality problems or preceding an important external audit.

External audits serve different purposes:

- Certification audits are performed by an official laboratory accreditation body (e.g., according to EN 45003 [61]) to prove the conformity of a product, a service, or a procedure with given standards. For example, a quality assurance system may be certified according to one of the ISO or EN standards. In the field of chemical analyses and measurements, however, accreditation has become more important than certification.
- Accreditation audits according to EN 45002 [60] are performed in order to obtain accreditation for all work performed by the laboratory or only for selected analytical procedures.
    - the permission to work in a special, legally defined field (so-called regulated field) or
    - the proof of qualified analytical work as a recommendation for clients (deliberate accreditation).

  Accreditation according to EN 45002 is valid in every EU member country.
- Assessment of performance by a client's auditor as part of a larger contract.
- Official audit of chemical or pharmaceutical manufacturer laboratories by a local or national authority in order to obtain permission for production or sales. This kind of audit takes place according to the OECD code on Good Laboratory Practice [165, 168, 169] or to the WHO Good Manufacturing Practice (GMP) Guide for Pharmaceutical Products [213] or the EU GMP code for medical devices [189] or to national regulations (e.g., FDA and EPA in the USA).

The procedure for an internal or external quality audit in an analytical laboratory can be based on different strategies:

- The process-oriented audit follows the chronology of work on a sample either forward, from sampling to report of results, or backward.
- The complaint-oriented audit reviews the most recent complaints and examines the handling of complaints and how information from complaints is used for quality improvement.
- The documentation-oriented audit checks all documentation and procedures according to the quality manual.
- The system-oriented audit checks the quality management system according to organograms, relationships, responsibilities, and flow charts for material and information.

4.3
Interlaboratory (or Round Robin) Tests

For an interlaboratory test, a reference laboratory dispatches partial samples of an evidently homogeneous original sample to the laboratories participating in the trial. These laboratories then carry out their analyses according to the provided instruction card and send the results back to the provider of the interlaboratory test.

Interlaboratory tests can be performed for different purposes; organization, examination, and evaluation of the results depend on the respective objective.

4.3.1
Interlaboratory Tests for Process Standardization

Interlaboratory tests performed in order to formulate an official standard method describe the *current* analytical status of the participating laboratories after a procedure has been established. They serve to determine the quality data of an analytical process: i.e., the characteristic process data such as recovery rate, comparative variation coefficient, and repeat variation coefficient. In this way, the reliability of an investigative method can be tested and documented.

The execution and evaluation of interlaboratory tests is standardized, for example, according to ISO 5725-2 [71]. Different outlier tests are applied, whereby a distinction is made between three types of outliers:

- laboratory-internal outliers (type 1)–Grubbs test,
- outliers in the laboratory mean (type 2)–Grubbs test,
- outliers in the laboratory precision (type 3)–Cochran test.

The elimination of outliers is legitimate since the standards should represent the mean performance of the laboratory. The characteristic process data published in the directory of official standard methods can therefore be used as a reference point for internal analytical quality assurance (AQA).

Since participants in interlaboratory tests for the determination of characteristic process data are almost always renowned research, industrial, and government laboratories, this may in some cases result in unrealistic data. For this reason, certain interest groups perform their own interlaboratory tests in order to establish the characteristic process data of individual parameters for all laboratories involved in common measurement programs.

4.3.2
Interlaboratory Tests as Proof of Laboratory Performance (Proficiency Test)

Interlaboratory tests are performed in order to test, and ensure in the future, the performance of participating laboratories and the comparability of their analytical results. Without comparable analytical results (from all participating laboratories assured over the entire time span), no meaningful description of an entity, a body of water for instance, can be provided.

The comparison of analytical results is even more important if the quality of results is doubted in a disagreement (e.g., for the calculation of sewage fees). In this case, the respective laboratory must be able to document that the required precision and accuracy was maintained on the day that the result was produced. This means that the laboratory must be included in an internal and external quality assurance system. The AQA of a laboratory is comprised of participation in interlaboratory tests.

In water analysis, for example, these interlaboratory programs have been organized at the state level for those laboratories which are active in government water management [157, 206]. The inspection of licensed laboratories occurs regularly, at least once a year.

In the past the execution and evaluation of interlaboratory tests for laboratory monitoring was carried out according to ISO 5725-2. Only the procedure for the elimination of outliers was different from the standard method:

No outliers of types 1, 2, or 3 were eliminated. Instead, previously agreed upon minimum requirements were prescribed for the evaluation of these tests, for example, a trueness of $\pm 10\%$ or an accuracy of $\pm 30\%$ [83]. These values had to be obtained by all participating laboratories. If the minimum requirements were repeatedly not attained, then a laboratory could lose its license for the performance of certain analyses. The following were used as reference values for the evaluation:

- for a trueness test: either the true value of the sample, if known, or the mean of all results,
- for a precision test: either an agreed upon percentage or the repeatability from the trial defined as the mean within-laboratory precision.

The effort of organizing and performing interlaboratory tests involving multiple analyses per sample is very large, given the large number of laboratories to be monitored and the multitude of parameters. The required execution of multiple analyses of a control sample also provides the opportunity to manipulate the precision (e.g., proper determination of the mean value but manipulative reporting of highly precise individual data points for the mean in question). For these reasons, the Youden method has been suggested as a simpler alternative for performing interlaboratory tests (see Section 4.3.5.2).

The evaluation of interlaboratory tests according to ISO 5725-2 assumes a normal distribution of the data. Since this condition is not usually met by the proficiency tests, robust evaluation procedures [46, 157] were developed. These are still usable with an outlier proportion of 33%. In the past few years, a robust evaluation of data has been established for cross-national interlaboratory tests for proficiency testing (see Section 4.3.5.3).

4.3.3
Other Interlaboratory Tests

Interlaboratory tests are not only carried out in the development of analytical procedures or the assessment of the analytical performance of laboratories, but also:

- to test analytical procedures for their practicality, especially if they are to be used in a new area of analysis or in different matrices,
- to determine the characteristics of a reference material.

4.3.4
Planning and Execution of Interlaboratory (or Round Robin) Tests

The principle of interlaboratory tests is that identical samples are analyzed by different laboratories independently of one another, each under *repeatable conditions*. When these results are evaluated, characteristic data are said to have been obtained under *comparable or reproducible conditions*.

The determination of reference values during interlaboratory tests will not be discussed here in further detail. Measurement procedures should be performed according to ISO guide 35 [126].

4.3.4.1 Quality Management System of the Provider of an Interlaboratory Test [46, 198]

The proper execution of interlaboratory tests assumes that the provider him- or herself implements a quality management system and documents the following in accordance with ISO IEC 17025 [76] in the quality management manual:

a) the name and address of the interlaboratory test provider as well as organizations involved, if necessary, including an organizational chart;
b) competence in the type of interlaboratory test with regard to the matrices and the analytical parameters in various test areas;
c) name of the person having overall responsibility and of his or her deputy, as well as the quality manager;
d) statements concerning the quality policy and quality objectives, as well as standard operating procedures for individual steps involved in the execution of the interlaboratory test and quality assurance measures;
e) statements ensuring the impartiality of the organizer and for the equal treatment of all interlaboratory test participants as well as third-party confidentiality;
f) type and objective of the interlaboratory test;
g) competencies and responsibilities of the personnel involved as well as a description of qualifications and further training;
h) description of the individual stages of planning, execution, and evaluation of an interlaboratory test (in standard operating procedures);
i) description of the general quality assurance procedures, which are valid for all interlaboratory tests (in standard operating procedures);
j) description of the special quality assurance procedures, which are valid for individual interlaboratory tests (in standard operating procedures);
k) methods for the determination of the assigned value (expected value);
l) methods for testing stability and homogeneity;
m) organization of transportation of subsamples (dispatch);
n) description of data collection and the statistical evaluation and assessment of interlaboratory test data;
o) methods of validating the evaluation software;

p) specification of pass criteria;
q) reporting to participants and specifications concerning the regular exchange of experiences and knowledge;
r) a system of documentation, document control, and archiving data;
s) corrective measures in the event of subsamples being mixed up, sample instability, and objectively false target values;
t) procedures for the handling of complaints, in particular questions of the subsequent correction of measurement data and failure to meet agreed deadlines;
u) decisions about the acceptance or refusal of laboratory results if delivery times are exceeded or changes are requested to previously submitted laboratory results;
v) methods of performing internal audits and reviews;
w) participation of the proficiency test provider in interlaboratory tests of other providers for the purpose of objectively assessing his or her own laboratory quality;
x) procedures for subcontract assignments;
y) the composition of a technical and scientific organization and assessment group.

4.3.4.2 Planning the Interlaboratory Test [46, 70, 71]

For the preparation of each interlaboratory test, a plan is devised and documented containing instructions concerning the following points:

a) description and objective of the interlaboratory test;
b) appointment of a test manager and a deputy;
c) sample matrix (type and origin, methodology for obtaining it, transport);
d) parameters to be analyzed, and if necessary, specifications of mandatory analytical procedures as well as concentration ranges to be covered in the interlaboratory test;
e) procedures for the specification or determination of assigned values (expected values);
f) number of different samples for each participant;
g) expected number of participants;
h) for spiked and synthetic samples:
   - proof of the traceability of the purity of the employed substances,
   - substances used to prepare the samples with specifications of the manufacturers/suppliers, the product number, the purity, as well as any pre-treatments that might be necessary,
   - method of preparing the stock solutions,
   - detailed descriptions of spiking procedures;
i) type of container used for samples or subsamples and necessary preliminary cleaning/conditioning;
j) necessary preliminary investigations to identify incompatibilities and interferences;

k) preservation of the samples;
l) quantity and type of sample sent to each participant (a larger quantity than necessary for the planned analyses, should there be a demand for an opportunity to "practice"?);
m) method of homogenization and homogeneity check;
n) timetable for all the phases of the interlaboratory test and method of ensuring adherence to these deadlines;
o) information for potential participants prior to registration;
p) information to be dispatched with the subsamples;
q) a system for ensuring confidentiality:
   - withholding the expected value from the participants,
   - withholding intermediate results,
   - protection of participant data,
   - measures to avoid collaboration between the participants;
r) procedure for the evaluation and assessment for proficiency testing by interlaboratory comparison: establishing accuracy boundaries;
s) organization of contact with the participants:
   - who may give what kind of information? (see also point q),
   - documentation of the contacts;
t) information that the participants will receive after the evaluation;
u) final publication of the interlaboratory test results.

4.3.4.3 Interlaboratory Test Samples

When selecting sample material, one should consider the objectives of the interlaboratory test, the target concentration levels, the necessary homogeneity and stability of the samples, as well as the logistics of transport and storage.

Different types of samples may be used for interlaboratory tests:

1. Natural samples that have not been preserved.
2. Preserved natural samples.
3. Natural samples, originally having a low concentration of the analyte, which have been spiked.
4. Synthetic samples with a problem-oriented matrix.
5. Synthetic samples without a problem-oriented matrix.

The "true value" (expected or assigned value) is only known for sample types 4 and 5. The mean analytical result, omitting outliers, is used primarily as the assigned value (conventional true value) for the remaining sample types [157]. Identical partial samples are taken from the interlaboratory test sample and are distributed to each of the participating laboratories.

4.3.5
Procedures for the Execution and Evaluation of Interlaboratory Tests

Interlaboratory tests are executed and evaluated according to the following standard methods:

- ISO 5725-2 [71] or DIN 38402 parts 41 & 42 [44, 45]
- Youden [217]
- ISO guide 43–1 [74] or ISO/CD 20612 [46]

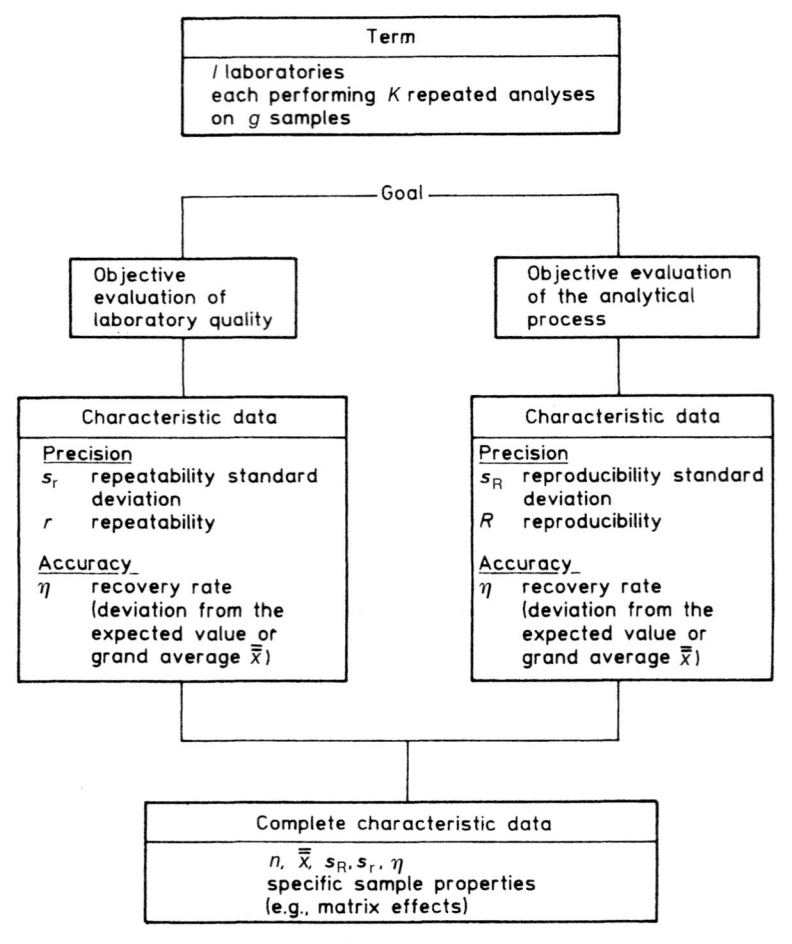

```
                         ┌─────────────────────────────────┐
                         │             Term                │
                         ├─────────────────────────────────┤
                         │ I laboratories                  │
                         │ each performing K repeated      │
                         │ analyses on g samples           │
                         └─────────────────────────────────┘
                                ────── Goal ──────
       ┌──────────────────────┐                    ┌──────────────────────┐
       │ Objective            │                    │ Objective evaluation │
       │ evaluation of        │                    │ of the analytical    │
       │ laboratory quality   │                    │ process              │
       └──────────────────────┘                    └──────────────────────┘

       Characteristic data                         Characteristic data

       Precision                                   Precision
       s_r   repeatability standard                s_R   reproducibility standard
             deviation                                   deviation
       r     repeatability                         R     reproducibility

       Accuracy                                    Accuracy
       η     recovery rate                         η     recovery rate
             (deviation from the                         (deviation from the
             expected value or                           expected value or
             grand average x̿)                            grand average x̿)

                         ┌─────────────────────────────────┐
                         │   Complete characteristic data  │
                         ├─────────────────────────────────┤
                         │ n, x̿, s_R, s_r, η               │
                         │ specific sample properties      │
                         │ (e.g., matrix effects)          │
                         └─────────────────────────────────┘
```

Fig. 4-1 Interlaboratory test evaluation according to ISO 5725-2 or DIN 38402 A-42; characteristic data.

4.3.5.1 Interlaboratory Test Programs According to ISO 5725–2

4.3.5.1.1 General Procedures

At least $l = 8$ laboratories, preferably $l > 15$, should participate in an interlaboratory test. Each of these laboratories normally performs $K = 4$ parallel determinations on each of the samples using the same, agreed upon analytical procedure. In special cases, there may be more or fewer parallel determinations, but never fewer than two.

The mean precision achieved by the individual laboratories (*repeatability standard deviation*, s_r) is determined on the basis of repeated determinations. Furthermore, it is possible to make an appraisal of the trueness for the individual laboratories by comparing the laboratory means with the "conventional true value" x_{exp} (expected value or grand average).

In order to assess the analytical process, the *reproducibility standard deviation* (s_R) is calculated for the precision statement and the *recovery rate* (η or RR) may be calculated for the trueness statement.

Figure 4-1 gives an overview of characteristic data investigated by an interlaboratory test according to ISO 5725-2.

The results of the interlaboratory test evaluation may be depicted graphically.

The center line represents the mean of the analytical results of all laboratories (not including outliers). The laboratory mean values ($\bar{x}_i$) and the respective standard deviations (s_i), and in some cases also the 2 s_i range (or $CI(\bar{x})$, $CI(x_i)$), for each individual laboratory are entered in a graph.

For interlaboratory tests used to standardize processes, type 2 (extreme mean □) and type 3 (imprecision *) outliers are marked with special symbols (see Figure 4-2).

For interlaboratory tests used for proficiency testing of laboratories, on the other hand, the limits of precision (e.g., 30%) are drawn on the graph and these are

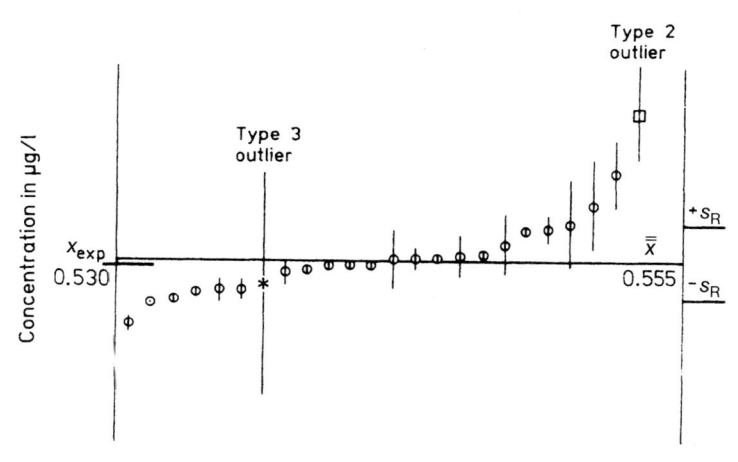

Fig. 4-2 Graphical representation of interlaboratory test results (process standardization).

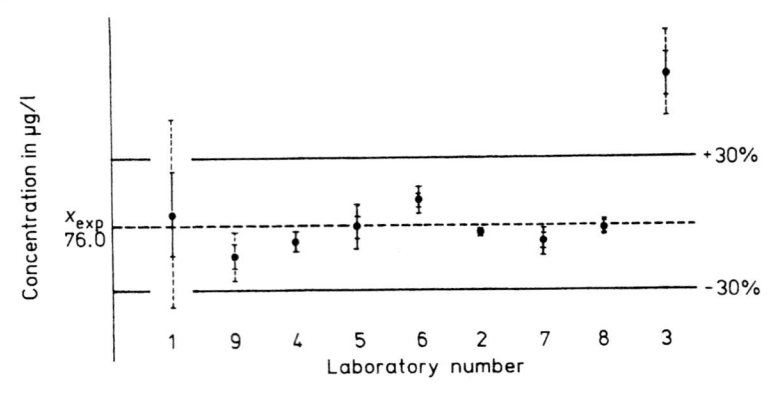

Fig. 4-3 Graphical representation of interlaboratory test results (laboratory monitoring), example.

used as further evaluation criteria (see Figure 4-3). It is therefore possible that type 2 or type 3 outliers also fulfill the quality requirements and that the laboratories in question would not lose their license for the performance of certain analyses.

4.3.5.1.2 Interpretation and Application of the Results

If an interlaboratory test is performed within the framework of process standardization, then the data obtained possibly form a basis for further measures such as:

– improvement of the analytical process in general,
– verification and improvement of the process in individual laboratories.

If such measures are necessary, then the interlaboratory test must perhaps be repeated. Only then can the performance of the analytical process in question be documented (e.g., as an official standard method).

The characteristics of so-called *validation interlaboratory tests*, which are documented in official standard methods, are of particular importance:

1. According to EN ISO IEC 17025, standardized analytical methods are considered validated; they indicate interlaboratory test characteristics which serve as orientation values, for example, for the attainable precision during the implementation of the analytical procedure (Phase II) in the routine.
2. The reproducibility standard deviation determined in the interlaboratory test can be consulted to estimate the measurement uncertainty (see Section 3.2.7).

Careful planning and execution of these interlaboratory tests is essential in order to obtain meaningful interlaboratory test data. For this reason, a guide to the execution of validation interlaboratory tests [27] was created for water analysis.

4.3.5.2 The Youden Method of Interlaboratory Tests [217, 218]

4.3.5.2.1 General
For laboratory monitoring, interlaboratory tests performed according to ISO 5725-2 [71] require much effort, especially because a very large sample volume must be provided by the organizer for $K = 4$ repeated analyses per laboratory.

The additional analysis effort for the laboratories being monitored is an important argument against performing interlaboratory tests in this way, since not only one, but perhaps many parameters of differing concentrations must be investigated at least once per year (more often if possible).

Therefore, Youden's method for the execution and evaluation of interlaboratory programs is suited to laboratory monitoring, since it is a simple statistical method for obtaining information about precision and systematic errors from analytical results without much effort. There is less additional burden on the laboratory since only two additional samples must be analyzed once. The interlaboratory test is evaluated graphically by plotting the results as a Youden plot [218]. However, applications of the Youden plot are not only limited to laboratory monitoring by means of interlaboratory tests.

4.3.5.2.2 Possible Applications of the Youden Plot

4.3.5.2.2.1 Application in External Analytical Quality Assurance
In external analytical quality assurance, the Youden plot may be used:

1. to compare analytical results, i.e., for monitoring the trueness (testing for the existence of systematic errors) and precision of individual laboratories [11, 16], and
2. to compare analytical processes, i.e., the determination of process suitability [218].

When comparing analytical processes, one can, for example, also detect systematic errors of the analytical process if the group of laboratories using the process achieves similar results. Prerequisites for this are a high precision and the absence of other systematic errors.

4.3.5.2.2.2 Application in Internal Analytical Quality Assurance
For laboratory internal analytical quality assurance, the Youden plot may be used:

1. for precision and trueness control, in the place of control charts [218],
2. to compare the performance of different operators,
3. to compare different analytical processes or different process variants and, therefore, for process optimization [218].

4.3.5.2.3 Evaluation Using the Youden Method

4.3.5.2.3.1 Direct Evaluation for Nearly Equal Standard Deviations
According to Youden, each of a sample pair of similar but not equal concentrations (sample A and sample B) are analyzed in each participating laboratory. The

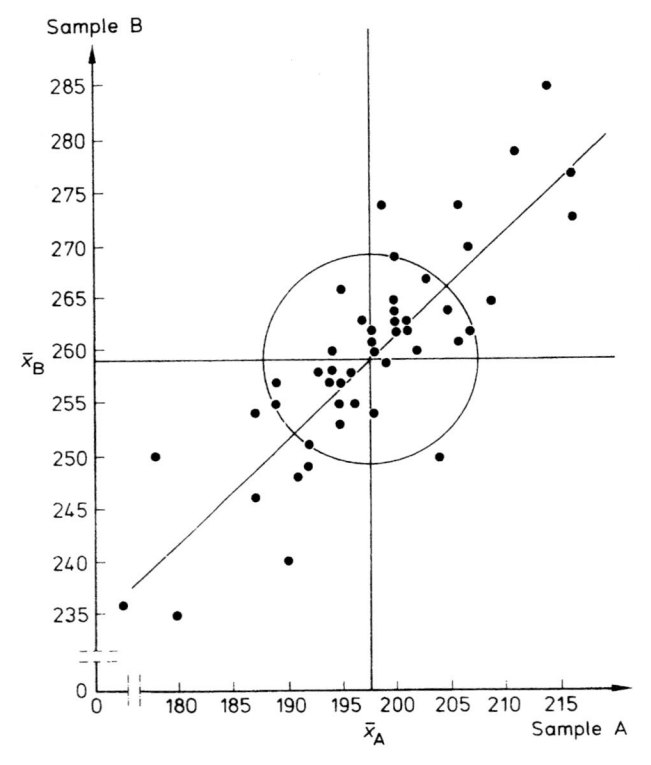

Fig. 4-4 Sample Youden plot, direct evaluation for roughly similar standard deviations.

results for sample A are plotted on the abscissa and those for sample B are plotted on the ordinate. The mean values $\bar{x}_A$ and $\bar{x}_B$ are calculated from the results for the respective samples and are plotted as lines parallel to the abscissa and ordinate, so that four quadrants appear (see Figure 4-4).

If a homogeneous distribution of the points in a circle around the intersection of the mean value lines is obtained, then primarily random errors exist.

On the other hand, a heterogeneous distribution, i.e., an accumulation of points in the upper-right and lower-left quadrants, indicates systematic errors.

In order to be able to better judge these systematic errors, Youden set out the following requirements. Samples should

- have the same type of matrix,
- have approximately the same concentration range,
- have approximately the same imprecision of the measured values,
- be analyzed using the same analytical process.

In this case, the division of both axes can be chosen to be identical. For such equal division of the axes, an additional 45° line may be drawn through the intersection of the mean value lines. Systematic errors can be detected if the points lie on or near the 45° line. The distance of a given point from the intersection of the mean value lines is a measure of the size of the systematic error (see Figure 4-4).

4.3.5.2.3.2 Direct Evaluation for Different Standard Deviations

When using different methods or different concentration ranges, one usually obtains different standard deviations of the analytical results. Therefore, the points entered do not lie within a circle but, if only random errors exist, within an ellipse (see Figure 4-5). The large axis has the same meaning as the 45° line in the Youden plot described in Section 4.3.5.2.3.1. However, this variation of the Youden plot provides no additional information. To simplify the evaluation, only two samples of the same type and with similar concentration and approximately the same standard deviation may be analyzed. The form of the normalized evaluation may be chosen in order to offset small differences in standard deviation.

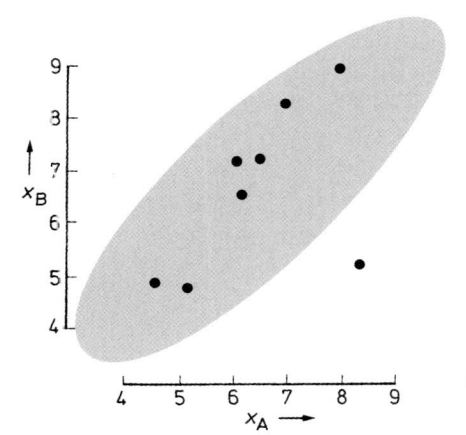

Fig. 4-5 Elliptical Youden plot.

4.3.5.2.3.3 Normalized Evaluation

A normalized Youden plot is used for the objective comparison of analytical results [218]. The mean values ($\bar{x}_A$ and $\bar{x}_B$) and standard deviations (s_A and s_B) are calculated for the analytical results of the respective samples:

$$\bar{x}_A = \frac{1}{N} \cdot \sum_{i=1}^{N} x_{Ai} \qquad \bar{x}_B = \frac{1}{N} \cdot \sum_{i=1}^{N} x_{Bi} \qquad (123)$$

$$s_B = \sqrt{\frac{1}{N-1} \sum (x_{Ai} - \bar{x}_A)^2} \qquad s_B = \sqrt{\frac{1}{N-1} \sum (x_{Bi} - \bar{x}_B)^2} \qquad (124)$$

The abscissa and ordinate are established and divided so that $\bar{x}_A$ lies in the center of the abscissa and $\bar{x}_B$ lies in the center of the ordinate. The distance between

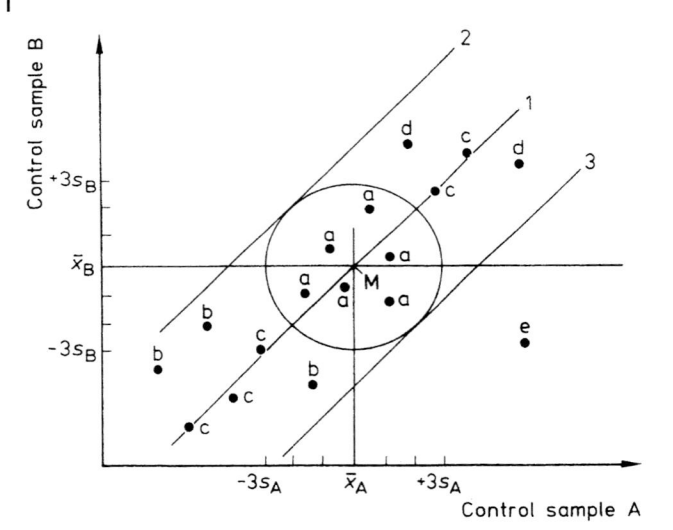

Fig. 4-6 Normalized Youden plot.

the origin and the center of the abscissa or ordinate should equal six times the standard deviation, s_A or s_B, respectively. The axes are scaled in units of the standard deviation, and the unit length for both axes is the same. The mean value lines are drawn parallel to the abscissa and ordinate at the level of the mean value. A circle with a radius of 3 units is drawn around the intersection of the mean value lines. This represents the 3s confidence interval, i.e., 99.7% of analytical result pairs can be expected to lie within this circle. In addition, a 45° line is drawn through the intersection of the mean value lines. Parallel to this line, two additional lines are drawn as tangents to the 3s circle. Lastly, the result pairs are added to the prepared Youden diagram as points (see Figure 4-6).

The coordinates of the points in relation to centerpoint M are calculated as follows:

$$\frac{x_{Ai} - \bar{x}_A}{s_A} \quad \text{and} \quad \frac{x_{Bi} - \bar{x}_B}{s_B} \tag{125}$$

Evaluation:

a) Points within the 3s circle: These results lie in the permissible confidence interval and are only subject to small (random) errors (points marked with *a* in Figure 4-6).
b) Points within the tangents but outside of the circle: Systematic errors exist for these results (points marked with *b, c,* and *d* in Figure 4-6).
c) Points on or near the 45° line and within the circle: the work that produced these results was very precise.
d) Points on or near the 45° line but outside of the circle: these results are indeed precise but are subject to a constant systematic error (points marked with *c* in Figure 4-6).

e) Points outside of the tangents: gross errors exist for these results (point marked with *e* in Figure 4-6).

If this fictitious example represented an interlaboratory comparison evaluation, then only the six laboratories marked with *a* would have performed good work. The rest would be outliers.

If interlaboratory tests are used for external monitoring of laboratories and if quality requirements are formulated for the analytical process being investigated, then the normalization of individual data (Eq. (125)) may be derived from the precision requirements: instead of s_A and s_B, the normalization would be performed using the required precision limit value, s_{req}.

Example:

Required precision: 20 %

$$20\% = \frac{s \cdot t}{\bar{x}} \cdot 100\%$$

This yields $s_{A_{req}}$ and $s_{B_{req}}$:

$$s_{A_{req}} = 0.2 \cdot \frac{\bar{x}_A}{t}$$

$$s_{B_{req}} = 0.2 \cdot \frac{\bar{x}_B}{t}$$

where $t = t\,(P = 95\%, f = \infty) = 1.96$

If the expected values or conventional true values of the samples are adequately known, then these values may be used instead of the mean values $\bar{x}_A$ and $\bar{x}_B$.

4.3.5.2.3.4 Simplified Evaluation

Sokolowski originated the simplified Youden plot [194]. This type of evaluation is advantageous because no identical sample types and concentration ranges are required; varying unit sizes may be chosen for the two standard deviations.

The mean value lines and the 45° line are also drawn on a simplified Youden plot, but a rectangle with side lengths $2 \cdot s_A$ and $2 \cdot s_B$, roughly marking the 95 % confidence intervals, is drawn instead of the 3s circle (see Figure 4-7).

Since the tangents to the 3s circle are omitted, part of the information is lost, and the crossover between gross and systematic errors becomes less well defined.

Interpretation of the simplified Youden plot

a) Points at the intersection of the mean value lines: the analytical results are correct and precise (point A).
b) Points within the rectangle: analytical results with random errors (points B).

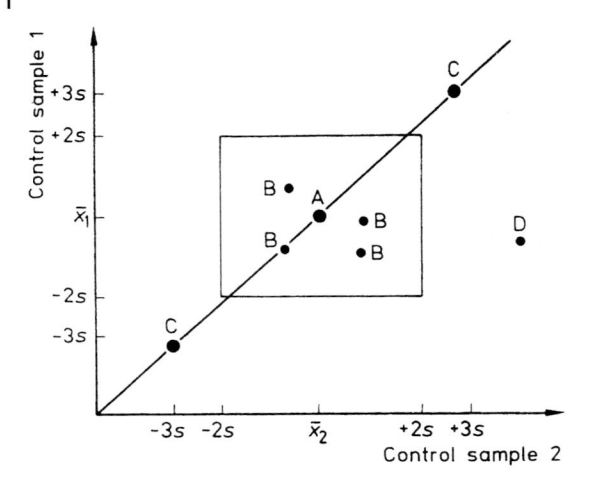

Fig. 4-7 Simplified Youden plot, example.

c) Points on or very near to the 45° line within the rectangle: analytical results with high precision.
d) Points on or very near to the 45° line outside of the rectangle: analytical results that are precise but subject to systematic errors (points C).
e) Points far away from the 45° line: analytical results with gross errors (point D).

Since its informational content is less than that of the normalized plot, a simplified Youden plot should only be used when dealing with very different sample types and concentrations.

4.3.5.2.3.5 Three- and Four-Dimensional Youden Plots

In Switzerland, interlaboratory test programs are customarily evaluated using the Youden method [16]. Since three or four samples are often analyzed for these programs, an expanded three- or four-dimensional Youden plot was developed, which can be constructed using a computer. For the three-dimensional evaluation, random errors appear as spherical point clouds around the center point of the space. Systematic errors are noticeable as point clouds along the spatial diagonal.

The Youden plot may, in principle, be expanded for any number of dimensions and concentrations. However, it is difficult to detect systematic errors when different concentration ranges are involved that require different measuring ranges [16]. Just a different number of significant figures of the reported results indicates a different accuracy! This would also be typical, for example, of incorrectly chosen measuring or working ranges. Therefore, as far as is possible, it is recommended that samples in the same concentration range are selected for interlaboratory programs.

4.3.5.2.4 Further Statistical Evaluation

The statistical evaluation possibilities described in this subsection may only be applied if the two samples are identical with respect to their composition (matrix) and the concentration of the analyte.

4.3.5.2.4.1 Estimation of the Laboratory-Internal Standard Deviation, s_w

$$s_w = \sqrt{\frac{\sum (d_i - \bar{d})^2}{2(N-1)}} \tag{126}$$

where

$$d_i = x_{Ai} - x_{Bi} \tag{127}$$

and

$$\bar{d} = \frac{\sum d_i}{N} \tag{128}$$

s_w is an estimate of the precision or the random errors.

4.3.5.2.4.2 Estimation of the Measuring Accuracy, s_d

$$s_d = \sqrt{\frac{\sum (T_i - \bar{T})^2}{2(N-1)}} \tag{129}$$

where

$$T_i = x_{Ai} + x_{Bi} \tag{130}$$

and

$$\bar{T} = \frac{\sum T_i}{N} \tag{131}$$

s_d represents the standard deviation resulting from both systematic and random errors.

4.3.5.2.4.3 F-Test for the Proof of Systematic Errors

To prove systematic errors, the test value TV is calculated

$$TV = \frac{s_d^2}{s_w^2} \tag{132}$$

Systematic errors exist if $TV > F(f_1 = f_2 = N-1, P = 99\%)$.

4.3.5.2.4.4 Estimation of the Standard Deviation Between Laboratories, s_b

$$s_b = \sqrt{\frac{s_d^2 - s_w^2}{2}} \tag{133}$$

s_b is therefore a measure of trueness. The calculation of this standard deviation is mainly carried out in the case of significant F-tests.

4.3.5.2.4.5 Estimation of the Total Standard Deviation, s_t

$$s_t = \sqrt{\frac{f_b s_b^2 + f_w s_w^2}{f_b + f_w}} \qquad (134)$$

where

$$f_b = N - 1 \qquad (135)$$

and

$$f_w = N \qquad (136)$$

4.3.5.2.4.6 Outlier Tests

Before calculating the mean and standard deviations for the construction of a Youden plot, it is recommended that outliers be eliminated. A suitable outlier test (e.g., Dixon test) should be applied to the analytical results x_{Ai} or x_{Bi}.

The Danish Water Quality Institute [164] recommends classifying as outliers those laboratories from which the analytical results show extreme random errors. The criterion for outliers is set as follows:

$$\left| d_s - \bar{d} \right| > 2 s_w \qquad (137)$$

where d_s represents the difference in analytical results, $x_{As} - x_{Bs}$, of the suspected outlier laboratory.

Elimination of outliers before constructing the Youden plot offers the advantage that the calculated standard deviations get smaller and provide for a stricter evaluation of the interlaboratory test results. This becomes unnecessary if normalization and evaluation are based on external precision requirements.

4.3.5.2.5 Advantages and Disadvantages of Interlaboratory Test Evaluation Using the Youden Method

Compared to interlaboratory test programs performed according to ISO 5725-2, programs performed according to the Youden method have the following *advantages*:

- less effort for organizers and participants,
- evaluation is simple,
- constant-systematic errors can be detected,
- the determination of comparative precision is simple,
- manipulation is less likely.

The *disadvantages* of the Youden evaluation method are:

- only one concentration level per plot can be monitored,
- laboratory-internal statistical data cannot be determined,
- useful results are only obtained if 30 or more laboratories participate,
- characteristic data of the analytical procedure are not determined.

Based on the advantages and disadvantages, Youden evaluation of interlaboratory test results is recommended above all for laboratory monitoring. In contrast, execution and evaluation of interlaboratory tests according to ISO 5725–2 is especially suited for process standardization.

4.3.5.2.6 Example of a Youden Interlaboratory Test Program
(Data taken from ref. [164])

Each of 30 participating laboratories was sent two samples of which the chloride content was to be investigated. Various analytical processes were allowed, e.g., photometric determination, ion chromatographic determination, automated analyses, etc. The analytical results determined by the individual laboratories, as well as the mean values and standard deviations calculated from these data, were compiled in Table 4-1. Using these data, a Youden plot was drawn (see Figure 4-8).

In this example, no outliers were eliminated before the evaluation. Hence, only Laboratory 11 was identified as an outlier from the Youden plot. If the mean values and standard deviations were calculated after elimination of Laboratory 11, then smaller standard deviations would have resulted and Laboratory 26 would have been spotted as an outlier too.

Table 4-1. Analytical results of interlaboratory test participants.

Laboratory i	Sample A x_{Ai} in mg/l	Sample B x_{Bi} in mg/l	Laboratory i	Sample A x_{Ai} in mg/l	Sample B x_{Bi} in mg/l
1	6.60	6.00	17	7.20	6.80
2	7.00	6.20	18	8.33	7.97
3	5.90	5.60	19	6.00	6.00
4	8.10	6.90	20	7.24	6.54
5	6.10	5.60	21	6.70	6.10
6	6.81	6.17	22	8.90	8.10
7	7.50	6.90	23	6.60	6.00
8	7.29	6.83	24	7.60	6.60
9	6.50	5.50	25	6.90	6.20
10	6.90	6.10	26	10.00	10.00
11	14.85	11.88	27	7.00	5.60
12	8.50	7.70	28	7.30	6.80
13	5.30	4.60	29	7.10	6.49
14	6.30	5.70	30	7.90	7.40
15	5.70	5.00	mean	7.37 mg/l	6.65 mg/l
16	7.00	6.36	standard deviation	1.72 mg/l	1.43 mg/l

Mean values and standard deviations after eliminating the data from Laboratory 11:
$\bar{x}_A = 7.11$ mg/l $\bar{x}_B = 6.47$ mg/l
$s_A = 0.9996$ mg/l $s_B = 1.054$ mg/l

It is noticeable that all laboratories participating in this interlaboratory test have worked very precisely, since all points in the Youden plot lie very near the angle bisect. The precision of laboratories 11 and 26 is also high, but systematic errors exist.

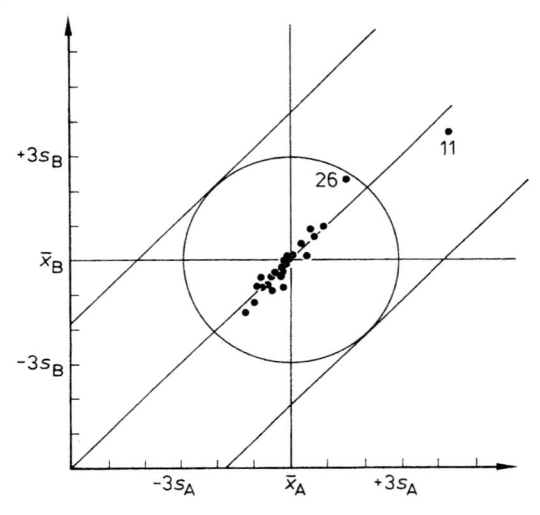

Fig. 4-8 Graphical evaluation of the interlaboratory test for chloride.

4.3.5.3 Interlaboratory Tests According to ISO Guide 43 [46, 74]

The German standard DIN 38402 part 45 (= ISO/CD 20612 [46]) has been created to guarantee uniform methods for the execution and evaluation of *proficiency tests*, and, if possible, for identical quality criteria to underlie the evaluation of laboratories. The standard is based on ISO guide 43-1 and -2, which sets out the framework for comparison tests, and describes the special details for the execution of proficiency testing by interlaboratory comparisons in water analysis.

4.3.5.3.1 General Procedures

Proficiency tests for investigation parameters, which are specified in relevant environmental laws and regulations, are arranged by different organizations. The German Federal Institute for Materials Research and Testing (BAM) regularly offers interlaboratory tests with regards to wastewater and polluted areas, among others. As for drinking water, the interlaboratory test providers in Germany, who are active in this area, have devised a concept to harmonize the execution of drinking water interlaboratory tests for the years 2004 to 2006 [97].

At least 12 laboratories should participate in interlaboratory tests according to DIN 38402 part 45. To get a more reliable picture of the efficacy of a laboratory, whenever possible, participants should analyze several samples with different concentration levels of individual parameters. Frequently, a series of different samples is prepared, each of which contains the parameters to be analyzed at different concentration levels. Each participant receives a set selection of subsamples of the different samples. This happens in order to effectively prevent the possibility of illegal arrangements. This also ensures a random distribution of different levels among the participants and a sense of equal treatment.

The proficiency tests should be executed under conditions that come as close as possible to those of routine operations. Therefore, the number of multiple determinations should correspond to those of routine measurements. The interlaboratory test provider can limit or prescribe the analytical procedures that can be applied depending upon the objective or the context of the interlaboratory test. If this is not the case, the participating laboratory may select the procedure freely.

4.3.5.3.2 Statistical Analysis
The evaluation of an interlaboratory test normally proceeds in four steps:

a) Definition of the Assigned Standard Deviation, s_{exp}
The assigned standard deviation, s_{exp}, serves as a measure for the calculation of the quality limits for the analytical results. It can be specified as a quality objective, though it is usually determined with the aid of statistical procedures from the analytical results of interlaboratory test participants. The Q-method, a procedure of robust statistics, is described in ref. [201] as a means of calculating the comparative standard deviation.

If for some reason the standard deviation calculated in this way is too wide or too narrow with regard to the quality objectives of analysis, then upper and lower limits for these standard deviations can be defined. If the calculated standard deviation exceeds or falls below one of these limits, the latter is defined as s_{exp}.

b) Definition of Expected Value, m_{exp}
The definition of the expected value is dependent on the type of sample origin (synthetic or natural samples). For the evaluation, the assigned concentration of a parameter (expected value) should, if possible, be based on measurements or confirmed by such. Taking into account the results of stability and homogeneity testing, the result uncertainty of the expected value must be clearly smaller than the total evaluation limits of the interlaboratory test for the respective analytical parameter.

In the event that the expected value is to be determined from the results of the participants, the HAMPEL estimator [162] can be applied as a robust statistical procedure.

c) Calculation and Optional Use of a Variance Function
In interlaboratory tests for assessing the proficiency of laboratories, samples of different concentrations with the same matrix are often distributed. An individual evaluation of these various samples frequently reveals fluctuations in the variances of the different concentration levels that might result in laboratories being unfairly assessed. To be able to correct for such fluctuations, the variances of the various samples can be defined with the aid of a variance function determined by a regression calculation based on the individual variances.

d) Calculation of Normalized Deviations of Results (Z Scores)

For the quality assessment of the results of laboratories, it is sensible to normalize the deviations of the results from the expected value, m_{exp}, using the standard deviation, s_{exp}. For this, Z scores are applicable [198].

A Z score can be calculated using the following equation:

$$Z = \frac{\text{analytical result} - m_{exp}}{s_{exp}} \tag{138}$$

Z scores result in tolerance limits in the assessment that are symmetrical with respect to the expected value, m_{exp}. In the case of a comparatively large standard deviation, s_{exp}, and measurements near the limit of quantification, this may result in the lower tolerance limit lying below the limit of quantification. If this happens, every laboratory whose analytical findings lie below the limit of quantification fulfills the quality criteria automatically.

Also, in the case of a smaller standard deviation, a clear preference for too small results is determined. In order to remedy the aforementioned disadvantages, modified Z_u scores are frequently used [46].

4.3.5.3.3 Interpretation and Application of Results

The interpretation of Z scores assumes a normal distribution of the analytical results. The probability that the absolute value of Z will not exceed the value 2 is then 0.9545, i.e., approximately 95%. The "quality limit" is therefore defined as the value $g = 2$. The quality criterion is considered fulfilled if the absolute value of Z does not exceed the value 2. Otherwise, with the probability of error $\alpha = 0.0455$, or approximately 5%, it can be assumed that the laboratory carried out the relevant analysis incorrectly. In individual cases, the limits 2.5 ($\alpha \approx 1\%$) or 3 ($\alpha \approx 0.3\%$) may be used in place of quality limit $g = 2$. The interlaboratory test provider decides this in light of the requirements.

For parameters that, in principle, cannot take negative values, Z_u scores are consulted as a quality criterion in place of Z scores.

The results of an interlaboratory comparison for proficiency test, on the other hand, are used specifically as a basis for decisions on the licensing of laboratories. According to ref. [157], for example, a laboratory has successfully participated in a proficiency test if at least 80% of all evaluated parameter sample level combinations in the interlaboratory test lie within the determined tolerance limits.

There are also other possibilities for interpretation:

- the monitored laboratories can rethink their internal quality assurance for certain parameters for which they have obtained poor results;
- plants, which are required to check their own waste water for example, can be informed about the analytical quality of the laboratories they contract to do the analytical work.

4.4
Effects of Internal Quality Assurance on the Results of Interlaboratory Tests

In 1982, a working group evaluating measurement methods for the "German Commission for a Pure River Rhine" was assigned the coordination of processes and performance data with regard to a given measurement task. The objective was the long-term assurance of the quality of analytical results. Using the sample parameter cadmium, the success of this interstate cooperation will be described.

The skill of participating laboratories was determined in the *initial interlaboratory test*. A standard solution of concentration 1.47 µg/l was distributed, which was to be analyzed by each laboratory. The results of this test (see Figure 4-9) showed that 9 of 24 laboratories did not achieve the required quality objective of 30 % precision set by EU directives [83]. Despite the simple matrix-free sample, only minimally comparable analytical results were achieved by the different laboratories.

The characteristic quantities of this interlaboratory test were determined (according to ISO 5725-2):

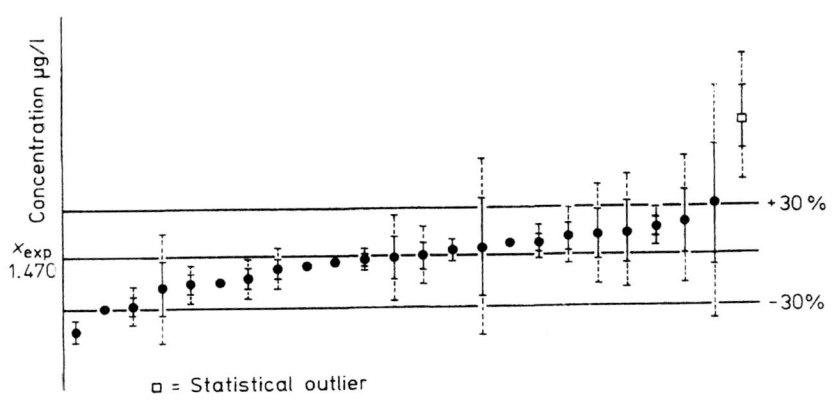

Fig. 4-9 Results of the first interlaboratory test, sample: standard solution 1.47 µg/l; with 2s scatter intervals (dotted lines) and 2s confidence intervals of the means (solid lines) of laboratory-internal results.

- *mean recovery rate*: $\overline{RR} = 97\%$
- *repeatability variation coefficient*: $CV_r = 7.6\%$
  (= a measurement of the mean laboratory internal precision achieved)
- *reproducibility variation coefficient*: $CV_R = 18.9\%$

Due to the poor results of this program, a working group was established which organized a priority program for laboratory-internal and -external quality assurance measures. This program included:

1. A comparison of the detailed descriptions of the laboratory-specific analytical procedures used.
2. The verification and improvement of the laboratory-internal process standard deviation, s_{xo}.
3. The management of control charts (mean value control charts, blank value control charts, RR charts).
4. The repetition of an interlaboratory test program.

Considering these points in turn:

For 1: The result of this comparison was that the instrumentation of the individual laboratories was for the most part similar. However, there were differences in sample preparation and the evaluation of measured values (calibration curve procedure or standard addition procedure). Gross or fundamental method errors could not be detected.

For 2 and 3: The laboratory-internal precision could be improved demonstrably during a "practice phase" and after the completion of necessary corrective measures.

For 4: After laboratory-internal improvement of the analytical quality, a second interlaboratory test was performed. A membrane-filtered surface water sample with a cadmium concentration of 0.67 µg/l was sent to the participating laboratories. The subsamples were analyzed four times using two different methods:

a) direct AAS determination,
b) AAS determination after wet chemical digestion according to the German (DIN) standard [65].

Results of the Second Interlaboratory Test

Figure 4-10 represents the results of the direct AAS determination. Only one laboratory (no. 5) did not achieve the accuracy requirements of ±30%. Furthermore, it should be noted that all laboratories which evaluated with calibration curves (marked with E in Figure 4-10) found results which were too low, while those laboratories which evaluated according to the method of standard addition (marked with A in Figure 4-10) obtained more accurate results. As a result, evaluation of the cadmium determination using the method of standard addition was recommended for subsequent tests.

The results of the AAS determination after wet-chemical digestion (see Figure 4-11) were similar to those of the direct AAS determination. In this case, the method of standard addition also provided more accurate results than the calibration curve method. The accuracy requirements were not achieved by three laboratories (nos. 1, 3, and 5).

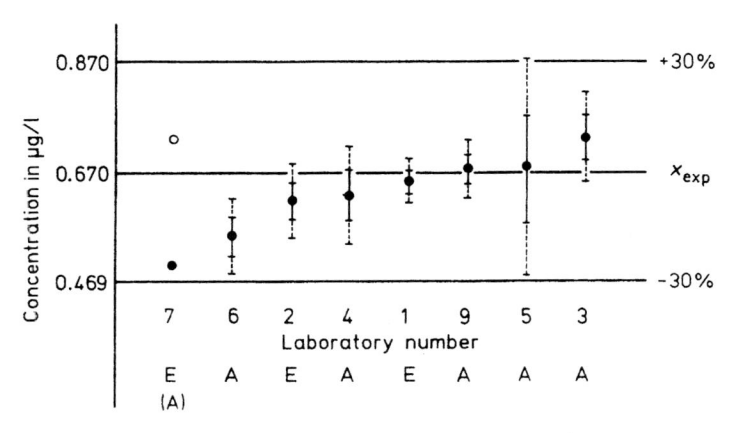

Fig. 4-10 Result of the second interlaboratory test, direct AAS determination.

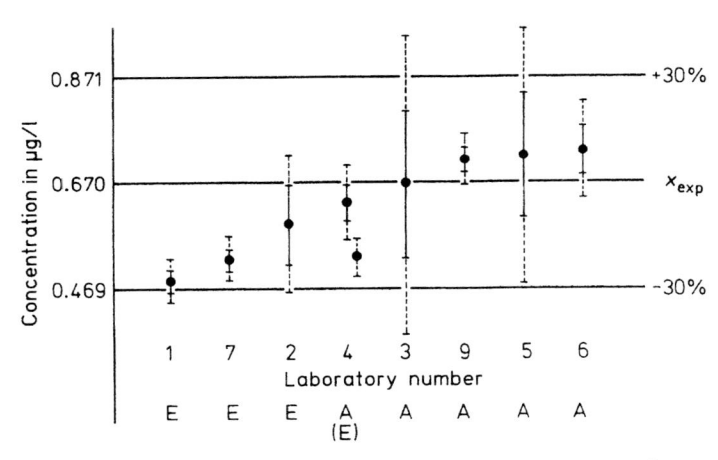

Fig. 4-11 Result of the second interlaboratory test, AAS determination after wet-chemical digestion.

The following characteristic quantities were obtained:

	Direct measurement	After wet-chemical digestion
$\bar{\bar{x}}$	0.637 µg/l	0.635 µg/l
$\overline{RR}$	95.1%	94.7%
CV_r	4.6%	7.2%
CV_R	11.4%	15.3%

Altogether, these results show that coordinated laboratory-internal quality assurance measures lead to a homogeneous and improved performance level for laboratories participating in an interlaboratory test program. The improvement in performance is even more pronounced if one takes into consideration that only standard solutions were analyzed in the first interlaboratory test whereas the second program utilized samples with an actual matrix and a lower concentration range.

4.5
Conclusion

Internal and external quality assurance should complement one another. In addition, communication between public and private analytical laboratories should not be limited to the reporting and discussion of interlaboratory test results. Especially in environmental analysis, regional quality working groups offer a guarantee for the comparability of measurement and analytical results, which is seen as an absolutely necessary prerequisite for the establishment of environmental databases and ecological studies.

Quality problems are neither arbitrary nor discriminating; they should be seen as a challenge for the improvement of the field of analytical chemistry. Or to use the words of Confucius:

> *He who makes a mistake and does not correct it*
> *is making another.*

5
Definitions

5.1
Quality and Quality Management

Quality [66]

Degree to which a set of inherent characteristics (i.e., of an intrinsic entity) fulfills given requirements.

Quality characteristic [66]

An inherent, constant characteristic of a product, process, or system related to a requirement.

Example: The most important quality characteristics of an "analytical method" are precision, trueness, analytical specificity, analytical selectivity, and analytical sensitivity.

Quality management [66]

Coordinated activities to direct and control an organization with regard to quality. Quality management (QM) usually covers:

- Quality planning: The part of QM focused on setting quality objectives and specifying necessary operational processes and related resources for fulfilling the quality objectives.

- Quality control: The part of QM focused on fulfilling quality requirements.

- Quality assurance: The part of QM focused on providing confidence that quality requirements will be fulfilled.

- Quality improvement: The part of QM focused on increasing the ability to fulfill the quality requirements.

Quality management system [66]

Management system to direct and control an organization with regard to quality.

Internal quality control [54]

Internal quality control encompasses all measures which are planned, ordered, and executed by a company laboratory itself.

External quality assessment [54]

External quality assessment includes all measures used by an independent body, i.e., one under a different administration, to check and advise a laboratory.

Accreditation

Through accreditation, an approved, external testing institution certifies that a laboratory possesses the ability to carry out certain analyses.

Certification

Through certification, an approved, external testing institution certifies that a product or process or service fulfills specific requirements, usually standards or compulsory regulations.

Statistical quality control [82]

That part of quality control in which statistical methods are used for planning, interpretation, and assessment.

Reliability [66]

Collective term used to describe the availability performance and its influencing factors: reliability performance, maintainability performance, and maintenance support performance.

Validation

Validating an analytical procedure determines whether the procedure fulfills the requirements of a specific purpose, including that the individual steps reliably fulfill all given specifications and quality characteristics.

Verification

Verification offers proof that specific requirements have been fulfilled.

5.2 Analytical Terms

Scheme for the terms analytical principle, analytical procedure, and analytical method:

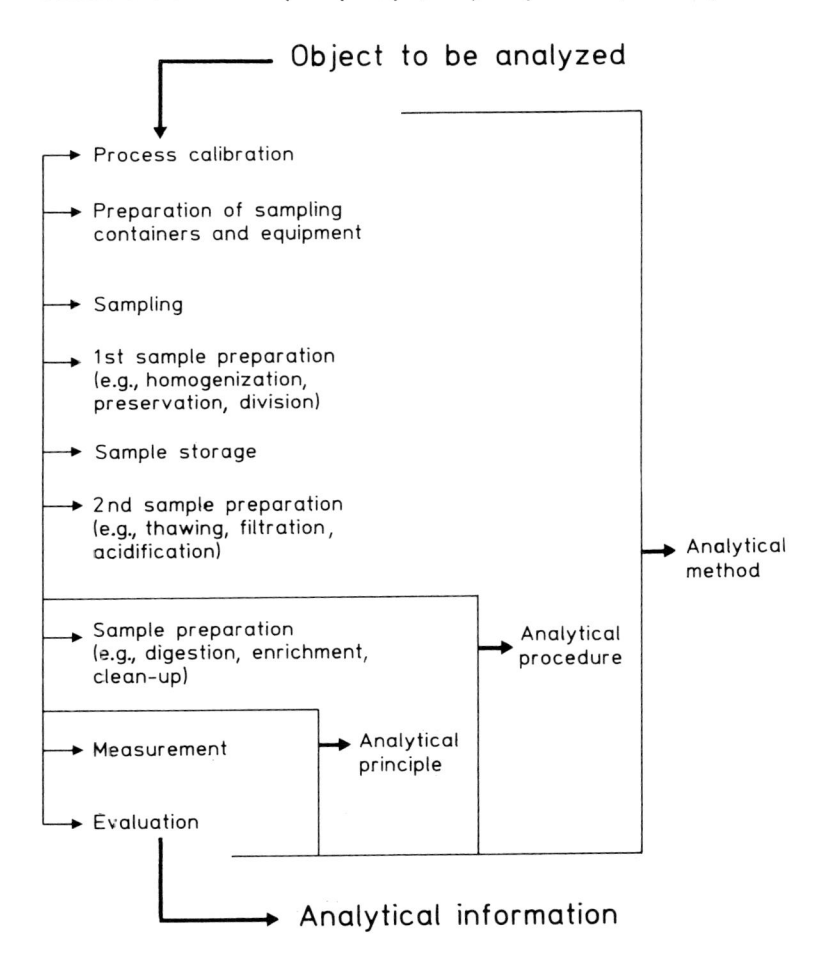

```
                    ┌──── Object to be analyzed
                    ↓
  → Process calibration
  → Preparation of sampling
    containers and equipment
  → Sampling
  → 1st sample preparation
    (e.g., homogenization,
    preservation, division)
  → Sample storage
  → 2nd sample preparation
    (e.g., thawing, filtration,
    acidification)                    → Analytical
                                        method
  → Sample preparation
    (e.g., digestion, enrichment,   Analytical
    clean-up)                       procedure
  → Measurement        → Analytical
                         principle
  → Evaluation
            → Analytical information
```

Measurement principles [30]

Characteristic physical properties which are used for measurement (e.g., photometry).

Measurement procedures [30]

Practical application of measurement principles:
- direct: obtaining measured values by immediate comparison with a reference quantity (e.g., volume measurements),

- indirect: deduction of measured values to different physical quantities (i.e., operational parameters).

Analytical principle

Refers only to measurements and the evaluation of measured values (see Scheme).

Analytical procedure

Refers only to the determination of a defined substance and the evaluation of measured values (see Scheme).

Analytical method

Combination of an analytical procedure with special (object-dependent) preparation techniques (sampling, sample preparation) (see Scheme).

Interlaboratory test (interlaboratory trial, collaborative trial, round robin test)

A number of laboratories, independently of one another, conduct a series of measurements or analyses for one or more characteristic values of samples of a specific material. Interlaboratory tests are carried out:

- as a validation of analytical methods,
- to determine the characteristics of reference materials,
- as a control of the efficiency of laboratories, "proficiency test".

Reference method

Reference methods are nationally or internationally recognized or prescribed analytical methods, which are described in an official document and for which results from interlaboratory tests are published for method validation. Reference methods follow particular analytical procedures:

- those described in an international guideline or standards manual (e.g., ISO, EN, AOAC),
- in a national standards manual (e.g., DIN, BSI, AFNOR, ...),
- in a national guideline manual or other national manual (e.g., VDI in Germany),
- specified or mutually agreed upon system procedures that have been prescribed as a result of legal regulation (e.g., national law, EU guidelines, etc.).

Synonyms for reference procedures are:

- standard methods,
- arbitration methods,
- reference methods.

Recovery rate

In a recovery experiment, a sample with known content, e.g., a (certified) reference material, spiked sample, etc., is analyzed. The relative yield is called the recovery rate (RR or η):

$$RR = \frac{\text{found content}}{\text{known content}} \cdot 100\%$$

or, in the case of spiking:

$$RR = \frac{\text{found (content after spiking} - \text{content before spiking)}}{\text{spiking quantity}} \cdot 100\%$$

Analytical specificity

Analytical specificity is the ability of an analytical method to register only the sought-after analytes in all relevant forms, whereby other components in or characteristics of the sample do not influence the analytical result.

Analytical selectivity

The ability of an analytical method to identify and quantify different co-analytes without reciprocal disturbance.

Analytical sensitivity [55]

Sensitivity describes the ability of an analytical method to differentiate between related values of an analyte (e.g., concentrations). It indicates to what extent the signal of the measuring system changes in relation to a change in the amount of the analyte and can be quantified using the slope of the calibration curve.

Resolution

Resolution is the smallest discernible difference between two readouts of an indicating device. In the case of a digital readout, this is the change of the indicated value, by which the last significant digit changes by one step.

Robustness, ruggedness

The robustness or ruggedness of an analytical procedure is the insensitivity of the performance characteristics in relation to changes in the analytical conditions. There is a distinction between:

- the "robustness" as the insensitivity in relation to real and distinctly definable changes of the procedural parameters – deviation from the analytical instructions – such as temperature or pH value, for example, and
- the "ruggedness" in the sense of comparability and the insensitivity in relation to a change in the implementing laboratory and/or staff and/or equipment.

Decision limit, x_{DL} [87]

The decision limit is the limit for the non-existence of a certain component with a probability of error α, for example, $\alpha = 5\%$: the probability that the component is still present is $\beta = 50\%$. The decision limit is also referred to as the "critical value".

Capability of detection, minimum detectable value, x_{MDV} [87]

The capability of detection indicates the minimum content that can be proven with a high, given probability. With a probability of error $\beta = \alpha$, the capability of detection is exactly twice as large as the decision limit. The capability of detection is also referred to as the "detection limit".

Limit of determination (limit of quantification), x_{LQ}

The limit of determination indicates the amount of substance content that can be determined with a given relative measurement uncertainty.

Measuring device

A measuring device is a piece of equipment which, alone or in conjunction with additional devices, determines and displays (i.e., indicates) the value of the measured variable.

Standard (measurement standard) [30]

A standard is a material measure (e.g., a 1 kg measurement standard), a reference measuring device, or reference material with the purpose of defining, representing, retaining, or reproducing a unit or one or more measurements. There is a distinction between international standards, national standards, reference standards, and usage standards.

A primary standard represents the highest technical measurement quality. Its value is accepted without reference to other standards.

The value of a secondary standard is assessed by comparison with a corresponding primary standard. Most certified reference materials are secondary standards because the certification of characteristic values is usually performed by means of a procedure involving a primary standard.

Traceability [30]

- Officially: the property of a measurement result or the value of a standard whereby it can be related to set references, usually international or national standards, through an unbroken chain of comparisons all having stated uncertainties.
- Colloquially: the possibility of being able to prove, on the basis of written documentation, the quality of all individual steps of an analysis and the resources used.

Calibration [30, 55]

The calibration of a system is the determination and establishment of a functional relationship between a numerical or measurable quantity and a concentration to be determined (activity, frequency, etc.) from data that are generally subject to random errors. An instrument is calibrated at defined intervals in order to verify its accuracy.

Adjustment [30, 55]

Adjustment is setting or aligning a measuring instrument so that measurement errors are kept as small as possible or that the total measurement error does not exceed the limits of error.

5.3 Analytical Results

True value [30, 52, 53]

Real (theoretical) characteristic value under the prevailing conditions during the analysis.

Note: The true value is often an ideal value because it can only be determined if all errors in the results can be avoided, or it is a result of theoretical considerations.

Conventional true value/assigned value [30, 53]

A value for the sake of comparison, the deviation of which from the true value is considered to be negligible for comparative purposes.

Note 1: The conventional true value is an approximation of the true value. It can be obtained from international, national, or usage standards, from reference materials, or standard methods (e.g., on the basis of specially organized experiments).

Note 2: There are several terms which are used as synonyms of "conventional true value". Examples are "expected value" and "target value". These terms are ambiguous and should be avoided. The latest ISO standards (e.g., ISO 13528:2005) use the term "assigned value".

Expected value [53]

The mean experimental result obtained using a continuously repeated experimental process under established conditions.

Actual value

An analytical result obtained momentarily, generally differentiated from the "expected value" (= conventional true value) by the presence of systematic and/or random errors.

Equivalency

The conventional view by which it is determined whether two analytical procedures for a certain applied purpose lead to comparable analytical results.

5.4
Deviation, Uncertainty

Accuracy [53, 70]

Accuracy is a qualitative term for the extent of the approximation of analytical results to the reference value, which may be either the true, real, or expected value depending on definition or agreement. Accuracy is therefore a generic term for precision and trueness.

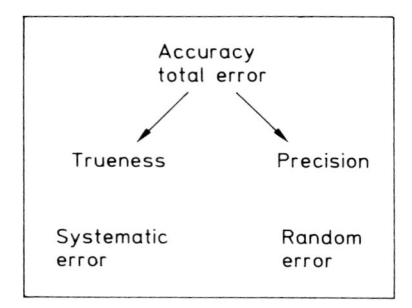

```
                Accuracy
              total error
              ↙         ↘
         Trueness       Precision

         Systematic     Random
         error          error
```

The determining factor for the overall error is the largest individual error.

The random error may be decreased by a factor $\frac{1}{\sqrt{N}}$ by repeating the analysis N times.

Systematic error may only be eliminated by the removal of its cause or by corrective calculations involving application of the recovery function.

Precision [53, 57, 70]

Precision is a qualitative term for the extent of the mutual approximation of analytical results, independent of one another, by repeated use of a set analytical method under defined conditions.

The conditions under which analytical results are obtained must be reported exactly. One should note the distinction between repeatability and reproducibility.

Repeatability, within-run (or batch) precision [55, 70]

Repeatability is a qualitative term for the extent of reciprocal approximation of analytical results under repeatability conditions. The terms "precision under repeatability conditions" and/or "serial precision" may also be used.

Repeatability conditions: Conditions for obtaining independent analytical results consisting of the repeated use of a set analytical process or method on the same object (same material, same sample) by the same operator in small time intervals (immediately following each other) with the same equipment (as well as identical supplies and other materials) in the same place (same laboratory) [53].

Reproducibility [55, 70]

Reproducibility is a qualitative term for the extent of reciprocal approximation of analytical results under reproducible conditions. The terms "precision under reproducible conditions" and/or "precision from lab to lab" are sometimes used in everyday speech.

Reproducibility conditions: Conditions for obtaining analytical results independent of each other consisting of the use of set analytical processes or methods on an identical object (same material, same sample) by different operators (investigators) with different equipment, supplies, and materials in different laboratories.

Between-run (or batch) precision [55]

Between-batch precision is a qualitative term for the extent of mutual approximation of analytical results obtained from determinations performed on the same material in the same laboratory, but in the course of different batches/series/days. Analyses within a series are carried out under repeatability conditions (see above).

Trueness, accuracy of the mean [53, 56, 70]

Trueness is a qualitative term for the extent to which the average value obtained from a large series of analytical results approximates the reference value, whereby this may be either the true value or the conventional true value according to definition or agreement.

Note 1: A quantitative measure of trueness is the systematic result deviation, also called "untrueness". This is determined from the difference between the experimentally derived mean value, obtained from several individual values, and the true value.

Note 2: For certain analyses, the conventional true value can also be established according to the state of the science.

Deviation, error [54]

An error is the difference between a characteristic value and the reference value of that characteristic. For quantitative characteristic values, this is equal to the difference between the analytical result and the reference value.

Total deviation, total error

The difference between the expected value and the actual value. The total deviation is comprised of the combined systematic and random errors.

Random deviation, random error [54, 122]

A component of the error which, in the course of a number of test results for the same characteristic, varies in an unpredictable way.

Note: Random errors cannot be eliminated by corrections. However, their influence on the result can be lessened by using a mean value obtained from several independent determinations.

Systematic deviation, systematic error [54, 122]

A component of the error which, in the course of a number of results for the same characteristic, remains constant or varies in a predictable way.

Note: For a defined process, systematic errors will be identical in extent and sign and reproducible. For quantitative characteristic values, the systematic error is equal to the difference between the expected value and the true or conventional true value (bias).

- *Constant systematic error, additive deviation (constant bias):*
  The amount of a systematic error is independent of the amount of the analytical result (e.g., "measurements were always 10 mg/l too low").

- *Proportional systematic error, multiplicative deviation (proportional bias):*
  The amount of a systematic error increases or decreases with the amount of the analytical result (e.g., "measurements were always 7% too high").

Gross error [54]

A gross error (or blunder) is an error which could have been easily avoided if proper procedures had been followed.

Measurement uncertainty [75, 90]

Due to further inevitable "errors" with each procedural step, repeating the entire analysis of a sample can lead to analytical results that are not identical, but that deviate from one another to a greater or lesser extent. Consequently, only a result interval, or confidence interval, in which the true analytical result lies with a definite probability of error α (usually 5% or 1%) may be indicated. The mean of this interval is also called the expected value. The half-width of this interval, i.e., the distance of the lower or upper interval limit from the mean value, is the measurement uncertainty. If this uncertainty is to be expressed as the standard deviation, it is also referred to as the *standard uncertainty*. Because the total uncertainty usually consists of several uncertainty components, the designation "combined measurement uncertainty" is used. According to the law of error propagation, the combined measurement uncertainty can be calculated from the individual standard uncertainties and can then be used as the standard deviation for the calculation of the confidence interval. The probability of error – with which the true analytical result may be found in the interval "mean value $\pm$ one standard deviation"

– is still more than approximately 32%. For a maximum probability of error of 5%, the interval – now called "expanded measurement uncertainty" – must at least be doubled; for $\alpha = 1\%$, the interval must be widened to the mean value $\pm$ three times the standard deviation.

5.5
Materials, Samples

Matrix [54]

The matrix of a material is the totality of all parts of the material and their chemical and physical properties, including mutual influences.

Reference material [54, 130]

A reference material is a material or substance of sufficient homogeneity, the property or properties of which can be defined so exactly that it may be used for the calibration of measuring instruments, the checking of results obtained from measuring, testing, and analytical processes, and for the characterization of substance properties. If the characteristic value (actual value $\pm$ measurement uncertainty) is obtained as the result of an interlaboratory test or through agreement with a qualified institution or assessor, then the term "consensus value" can be used.

A certified reference material (CRM) is provided with a certificate indicating the uncertainty and the associated confidence level of one or more characteristic values that have been certified as a result of a determination procedure, and with which the traceability of the values will be attained from an exact realization.

CRMs are generally produced in batches, the characteristic values of which are determined within specified uncertainty limits through measurements on samples and are representative of the whole batch. CRMs fulfill the definition of "standards" (see above) in the "International Vocabulary of Metrology (VIM)".

Bulk sample, original sample, specimen

A bulk sample, also called a primary sample [63], is one taken on the spot, the content of a substance in which is to be determined quantitatively. The analytical result, x, therefore represents an approximation of the true value.

Analytical sample

An analytical sample is a sample which is obtained from the bulk sample after a work-up (e.g., disintegration, extraction, etc.) and, if necessary, dilution or concentration, and which is then used for the actual measuring process. An analytical sample is therefore identical to a bulk sample if the original sample may be used directly in the measuring process without work-up or dilution or concentration.

Measurement sample

A measurement sample is a sample for which the analyte content may be measured directly. A measurement sample is normally obtained from an analytical sample by the addition of a series of reagents. Measurement samples and analytical samples are therefore identical when no reagents are added to the analytical sample.

5.6
Statistical Tests

Confidence level, level of significance, P

The level of significance denotes the relative safety of a "statistical decision". If the underlying experiment for the decision were to be repeated very often (indefinitely), then the same decision would result in P percent of all cases. The confidence level can be calculated from the probability of error α: $P = 1 - \alpha$.

Probability of error, α

The probability of error denotes the probability that a decision is false. If the underlying experiment for the decision were to be repeated very often (indefinitely), then the decision must be revised in α percent of all cases.

Statistical test

A statistical test serves to clarify whether a discrepancy, i.e., a difference between data, is just random or meaningfully "significant". The decision can, however, only be made with a specific degree of certainty (level of significance, P), i.e., probability of error $\alpha = (1 - P)$.

Expected value t-test

A test of the statistical difference between an expected value (conventional true value) and a mean value $\bar{x}$ (normal distribution assumed or required) determined by means of N analyses. A test value, TV, is calculated and compared with the threshold value, t, from the t-table:

$$TV = \left| \frac{\bar{x} - x_{\exp}}{s} \right| \cdot \sqrt{N}$$

where $\bar{x} = \dfrac{1}{N} \sum\limits_{i=1}^{N} x_i$, $x_{\exp}$ = expected value, and $s = \sqrt{\dfrac{\sum\limits_{i=1}^{N}(x_i - \bar{x})^2}{N-1}}$

Decision:

$$TV \leq t\,(f, P = 95\,\%): \text{random difference}$$
$$t\,(f, P = 95\,\%) < TV \leq t\,(f, P = 99\,\%): \text{probable difference}$$
$$TV > t\,(f, P = 99\,\%): \text{significant difference}$$

$t\,(f, P)$ see Appendix A2.1.

Mean *t*-test

A test of the statistical difference between two means ($\bar{x}_1, \bar{x}_2$) obtained from two independent series of analyses (normal distribution assumed or required). A test value, *TV*, is calculated and compared with the threshold value, *t*, from the *t*-table:

$$TV = \left|\frac{\bar{x}_1 - \bar{x}_2}{s_d}\right| \cdot \sqrt{\frac{N_1 \cdot N_2}{N_1 + N_2}}$$

where $s_d = \sqrt{\dfrac{(N_1 - 1) \cdot s_1^2 + (N_2 - 1) \cdot s_2^2}{N_1 + N_2 - 2}}$

(for $N_1 = N_2 = N, f_1 = f_2; f = 2N - 2$ is valid)

Decision:

$$TV \leq t\,(f, P = 95\,\%): \text{random difference}$$
$$t\,(f, P = 95\,\%) < TV \leq t\,(f, P = 99\,\%): \text{probable difference}$$
$$TV > t\,(f, P = 99\,\%): \text{significant difference}$$

$t\,(f, P)$ see Appendix A2.1.

Variance *F*-test

The *F*-test is used to test the inequality of two variances, determined from two independent data series.

$$TV = \frac{\text{larger variance}}{\text{smaller variance}}$$

$$TV = \frac{s_1^2}{s_2^2}$$

Decision:

$$TV \leq F\,(f_1, f_2, P = 95\,\%): \text{random difference}$$
$$F\,(f_1, f_2, P = 95\,\%) < TV \leq F\,(f_1, f_2, P = 99\,\%): \text{probable difference}$$
$$TV > F\,(f_1, f_2, P = 99\,\%): \text{significant difference}$$

$t\,(f_1, f_2, P)$ see Appendix A2.2.

Grubbs test for outliers

To identify individual outliers, the mean value, $\bar{x}$, and the standard deviation, s, are calculated from the analysis data under repeatability or reproducibility conditions. The analysis value x^* showing the greatest difference from the mean is tested according to the following conditional equation:

$$TV = \frac{|x^* - \bar{x}|}{s}$$

Decision:

$$TV \leq rM\,(N,P = 90\%): \text{random difference}$$
$$rM\,(N,P = 90\%) < TV \leq rM\,(N,P = 95\%): \text{probable difference}$$
$$TV > rM\,(N,P = 95\%): \text{significant difference}$$

$rM\,(N,P)$ see Appendix A2.3.

6
References

1 Amador, E.: Quality Control by the Reference Sample Method, *Am. J. Clin. Path.* 50 (1968), 360.
2 AOAC: Interlaboratory Collaborative Study, Appendix D: Guidelines for Collaborative Study Procedures to Validate Characteristics of a Method of Analysis, AOAC, 2002.
3 ASTM (American Society for Testing and Materials): Standard Methods for the Examination of Water and Wastewater, 14th Edition (1975), 104 B, 26–33.
4 ASTM STP 15D: Manual on Presentation of Data and Control Chart Analysis, American Society for Testing and Materials (1976).
5 Barnard, A.; Mitchell, R.; Wolf, G.: Good Analytical Practices in Quality Control, *Analytical Chemistry* 50(12) (1978), 1079A–1086A.
6 Barnard, G. A.: Control Charts and Stochastic Processes, *J. R. Statist. Soc.*, B 21 (1959), 239–271.
7 Barwick, V. J.; Ellison, S. L. R.: (VAM Project 3.2.1 Development and harmonisation of measurement uncertainty principles, part (d)): Protocol for uncertainty evaluation from validation data (January 2000).
8 Barwick, V. J.; Ellison, S. L. R.: Measurement Uncertainty: Approaches to the Evaluation of Uncertainties Associated with Recovery, *Analyst* 124 (1999), 981–990.
9 Publication of a consensus document of the Federal States Working Group (Bund-Länder-Arbeitsgruppe) for Good Laboratory Practice on the topic „Gute Laborpraxis (GLP) und Datenverarbeitung". Bundesministerium für Umwelt, Naturschutz und Reaktorsicherheit, 28th October, 1996.
10 Publication of a consensus document of the Federal States Working Group (Bund-Länder-Arbeitsgruppe) for Good Laboratory Practice for archiving and storage of documentation and materials. Bundesministerium für Umwelt, Naturschutz und Reaktorsicherheit, 5th May, 1998.
11 Beyer, W. H. (Ed.): Handbook of Tables for Probability and Statistics, CRC Press, Boca Raton, Florida (1981).
12 Boehringer Mannheim, Wissenschaftliche Abteilung Diagnostika: Qualitätskontrolle Fehlersuche (1986).
13 Boroviczèny, K. G. von; Merten, R., Merten, U. P. (Eds.): Qualitätssicherung im Medizinischen Labor, Springer-Verlag, Berlin, Heidelberg (1987).
14 Burr, I. W.: The Effect of Non-Normality on Constants for $\bar{x}$ and R Charts, *Industrial Quality Control* (May 1967), 563.
15 Caulcutt, R.; Boddy, R.: Statistics for Analytical Chemists, Chapman & Hall, London, New York (1983).
16 Chatterjee, S., Price, B.: Regression Analysis by Example, J. Wiley & Sons, New York (1977).
17 Cheeseman, R. V.; Wilson, A. L.: Manual on analytical quality control for the water industry, Water Research Centre, technical report, TR 66, Medmenham (1978).
18 CITAC/EURACHEM Guide: Guide to Quality in Analytical Chemistry, An Aid to Accreditation (2002).
19 Council Directive 87/18/EEC, 18th December, 1986, on the Harmonization of

Quality Assurance in Analytical Chemistry: Applications in Environmental, Food, and Materials Analysis, Biotechnology, and Medical Engineering, Second Edition. W. Funk, V. Dammann, G. Donnevert
Copyright © 2007 WILEY-VCH Verlag GmbH & Co. KGaA, Weinheim
ISBN: 978-3-527-31114-9

Laws, Regulations and Administrative Provisions Relating to the Application of the Principles of Good Laboratory Practice and the Verification of their Applications for Tests on Chemical Substances, *Official Journal* L 015, 17/01/1987, p. 0029–0030.

20 Council Directive 89/569/EEC, 28th July 1989, on the Acceptance by the European Economic Community of an OECD Decision/Recommendation on Compliance with the Principles of Good Laboratory Practice, *Official Journal* L 315, 28/10/1989, p. 0001–0032.

21 Crosby, N. T.; Day, J. A.; Hardcastle, W. A.; Holcombe, D. G.; Treble, R. C.: Quality in the Analytical Chemistry Laboratory, ACOL, J. Wiley & Sons, Chichester, New York, Brisbane, Toronto, Singapore (1995).

22 Dammann, V.; Funk, W.; Marcard, G. v.; Papke, G.; Rinne, D.: Zur Problematik der Bestimmungsgrenze in der Wasseranalytik, *Vom Wasser* 66 (1986), 97–109.

23 Deutsche Gesellschaft für Qualität, Arbeitsgruppe 7: Qualitätsregelkarten, DGQ-Schrift Nr. 16–30, Beuth Verlag, Berlin and Cologne, 3rd Edition (1979).

24 Deutscher Akkreditierungsrat: Anforderungen an Prüflaboratorien und Akkreditierungsstellen bezüglich der Messunsicherheitsabschätzung nach ISO/IEC 17025, DAR-4-INF-08 (2001).

25 Deutscher Akkreditierungsrat: Vorstellung eines Konzepts zur Messunsicherheit im Prüfwesen in Verbindung mit der Anwendung der ISO/IEC 17025, DAR-4-INF-09 (2002).

26 Deutscher Kalibrierdienst: Angabe der Messsicherheit bei Kalibrierungen, DKD-3 Ausgabe 01/1998.

27 DEV A0–3 Strategien für die Wasseranalytik: Anleitung zur Durchführung von Ringversuchen zur Validierung von Analysenverfahren, DEV 57. Lieferung 2003.

28 Dewey, D. J.: Within-laboratory analytical quality control (Water Research Centre), Nyt Miljoestyr. Referencelab., Vandkvalitetsinst. (1979), 7:79, 6–17.

29 Dharan, M.: Total Quality Control in the Clinical Laboratory, The C. V. Mosby Company, St. Louis (1977).

30 DIN 1319 Teil 1: Grundlagen der Meßtechnik, Grundbegriffe (January 1995).

31 DIN 1333 Zahlenangaben (February 1992).

32 DIN 31051: Grundlagen der Instandhaltung, Begriffe und Maßnahmen (June 2003).

33 DIN 32633 Chemische Analytik – Verfahren der Standardaddition; Verfahren, Auswertung (December 1998).

34 DIN 32645: Nachweis-, Erfassungs- und Bestimmungsgrenze; Ermittlung unter Wiederholbedingungen; Begriffe, Verfahren, Auswertung (1994).

35 DIN 38402 Teil 11: Allgemeine Angaben – Probenahme von Abwasser (December 1995).

36 DIN 38402 Teil 12: Allgemeine Angaben – Probenahme aus stehenden Gewässern (June 1985).

37 DIN 38402 Teil 13: Allgemeine Angaben – Probenahme aus Grundwasserleitern (December 1985).

38 DIN 38402 Teil 14: Allgemeine Angaben – Probenahme von Rohwasser und Trinkwasser (March 1986).

39 DIN 38402 Teil 15: Allgemeine Angaben – Probenahme aus Fließgewässern (July 1986).

40 DIN 38402 Teil 16: Allgemeine Angaben – Probenahme aus dem Meer (August 1987).

41 DIN 38402 Teil 19: Allgemeine Angaben – Probenahme von Schwimm- u. Badebeckenwasser (April 1988).

42 DIN 38402 Teil 20: Allgemeine Angaben – Probenahme aus Tidegewässern (August 1987).

43 DIN 38402 Teil 30: Allgemeine Angaben – Vorbehandlung, Teilung und Homogenisierung heterogener Wasserproben (July 1998).

44 DIN 38402 Teil 41: Allgemeine Angaben – Ringversuche, Planung und Organisation (May 1984).

45 DIN 38402 Teil 42: Allgemeine Angaben – Ringversuche, Auswertung (September 2005).

46 ISO/CD 20612 (2005): Water quality – Interlaboratory comparisons for proficiency test of laboratories.

47 ISO 8466–1:1990, Water quality – Calibration and evaluation of analytical

methods and estimation of performance characteristics – Part 1: Statistical evaluation of the linear calibration function.
48 DIN 38402 Teil 71: Gleichwertigkeit von zwei Analyseverfahren aufgrund des Vergleiches von Analyseergebnissen und deren statistischer Auswertung (November 2002).
49 DIN 38406 Teil 1: Kationen – Bestimmung von Eisen (May 1983).
50 DIN 38407 Teil 2: Gaschromatographische Bestimmung von schwerflüchtigen Halogenkohlenwasserstoffen (February 1993).
51 DIN 38409–41, Ausgabe: 1980–12: Deutsche Einheitsverfahren zur Wasser-, Abwasser- und Schlammuntersuchung; Summarische Wirkungs- und Stoffkenngrößen (Gruppe H); Bestimmung Chemischen Sauerstoffbedarfs (CSB) im Bereich über 15 mg/l (H 41).
52 DIN 55350 Teil 12: Begriffe der Qualitätssicherung und Statistik, Merkmalsbezogene Begriffe (March 1989).
53 DIN 55350 Teil 13: Begriffe der Qualitätssicherung und Statistik, Begriffe zur Genauigkeit von Ermittlungsverfahren und Ermittlungsergebnissen (July 1987).
54 DIN 58936 Teil 1: Qualitätssicherung in der Laboratoriumsmedizin, Grundbegriffe (October 2000).
55 DIN 58936 Teil 2: Qualitätssicherung in der Laboratoriumsmedizin, Begriffe zur Qualität und Anwendung von Untersuchungsverfahren (June 2001).
56 DIN 58936 Teil 5: Qualitätssicherung in der Laboratoriumsmedizin, Kontrollkarten; Begriffe, Allgemeine Anforderungen (July 1983).
57 DIN 58937 Teil 4: Allgemeine Laboratoriumsmedizin, Anforderungen an die Beschreibung von Methoden (November 1988).
58 EN 14136:2004, Use of external quality assessment schemes in the assessment of the performance of in vitro diagnostic examination procedures.
59 DIN EN 1485, Ausgabe: 1996–11: Wasserbeschaffenheit – Bestimmung adsorbierbarer organisch gebundener Halogene (AOX); Deutsche Fassung EN 1485:1996.
60 EN 45002: General criteria for the assessment of testing laboratories (1989).
61 ISO/IEC Guide 58:1993, Calibration and testing laboratory accreditation systems – General requirements for operation and recognition (EN 45003).
62 ISO/IEC Guide 62:1996, General requirements for bodies operating assessment and certification/registration of quality systems (EN 45012).
63 EN ISO 15189:2003, Medical laboratories – Particular requirements for quality and competence.
64 EN ISO 19011:2002, Guidelines for quality and/or environmental management systems auditing.
65 EN ISO 5961:1994, Water quality – Determination of cadmium by atomic absorption spectrometry.
66 ISO 9000:2005 Quality management systems – Fundamentals and vocabulary.
67 EN ISO 9001:2000, Quality management systems – Requirements.
68 EN ISO 9004:2000, Quality management systems – Guidelines for performance improvements.
69 ISO 5479:1997, Statistical interpretation of data – Tests for departure from the normal distribution.
70 ISO 5725–1:1994, Accuracy (trueness and precision) of measurement methods and results – Part 1: General principles and definitions.
71 ISO 5725–2:1994, Accuracy (trueness and precision) of measurement methods and results – Part 2: Basic method for the determination of repeatability and reproducibility of a standard measurement method.
72 ISO 8466–2:2001, Water quality – Calibration and evaluation of analytical methods and estimation of performance characteristics – Part 2: Calibration strategy for nonlinear second-order calibration functions.
73 DIN V 38402 Teil 17: Allgemeine Angaben – Probenahme von fallenden, nassen Niederschlägen in flüssigem Aggregatzustand (May 1988).
74 ISO/IEC Guide 43–1:1997, Proficiency testing by interlaboratory comparisons – Part 1: Development and operation of proficiency testing schemes.

75 ENV 13005:1999, Guide to the expression of uncertainty in measurement.
76 EN/ISO/IEC 17025:2005, General requirements for the competence of testing and calibration laboratories.
77 Doerffel, K.; Eckschlager, K.: Optimale Strategien in der Analytik, Verlag Harri Deutsch, Thun, Frankfurt/M. (1981).
78 Doerffel, K.: Statistik in der analytischen Chemie, Wiley-VCH, Weinheim (2002).
79 Draper, N., Smith, H.: Applied Regression Analysis, Wiley Interscience, New York, Chichester, Brisbane, Toronto (1998).
80 Driscoll, J. L.; Gudzinowicz, B. J.; Martin, H. F.: Instrument Evaluation in Biomedical Sciences, Marcel Dekker Inc., New York, Basel (1984).
81 Duncan, A. J.: Quality Control and Industrial Statistics, R. D. Irwin Inc., Homewood, Illinois (1965).
82 E DIN 55350 Teil 11: Begriffe zu Qualitätsmanagement und Statistik – Teil 11: Begriffe des Qualitätsmanagements, Ergänzung zu DIN EN ISO 9000:2000–12.
83 Council Directive of 9th October 1979 concerning the methods of measurement and frequencies of sampling and analysis of surface water intended for the abstraction of drinking water in the Member States (79/869/EEC).
84 Eichordnung vom 12. Aug. 1988, Allgemeine Vorschriften, Physikalisch-technische Bundesanstalt, Deutscher Eichverlag Braunschweig, Bestell-Nr. 9560 und Anlagen, z. B. Anlage 12 zur Eichordnung: Volumenmeßgeräte für Laboratoriumszwecke.
85 Eisenhart, Ch.: William John Youden, 1900–1971, *Journal of Quality Technology*, Vol. 4, No. 1 (Jan. 1972), 3–6.
86 El-Nageh, M. M.: Basics of Quality Assurance for Intermediate and Peripheral Laboratories, WHO Regional Publications, Eastern Mediterranean Series, No. 2, 1992.
87 2002/657/EC: Commission Decision of 14th August 2002 implementing Council Directive 96/23/EC concerning the performance of analytical methods and the interpretation of results. *Official Journal* L 221, 17/08/2002, p. 8, corr. L 239, 06/09/2002, p. 66.
88 2002/657/EC: Commission Decision of 12th August 2002 implementing Council Directive 96/23/EC concerning the performance of analytical methods and the interpretation of results. *Official Journal* I 221, 17/08/2002, p. 8, corr. L 239, 06/09/2002, p. 8–16.
89 EPA (Environmental Protection Agency): Handbook for Analytical Quality Control in Water and Waste Water Laboratories, EPA-600/4–79-019, Office of research and development, Cincinnati, Ohio 45268 (1979).
90 EURACHEM/CITAC Guide: Quantifying Uncertainty in Analytical Measurement, 2nd Edition (2000).
91 EURACHEM Guide on Selection, Use and Interpretation of Proficiency Testing (PT) Schemes by Laboratories (2000).
92 EURACHEM Guide: The Fitness for Purpose of Analytical Methods – A Laboratory Guide to Method Validation and Related Topics (1st Edition, 1998).
93 EUROLAB Technical Report No. 1/2002: Measurement Uncertainty in Testing (2002).
94 European co-operation for accreditation: EA guidelines on the expression of uncertainty in quantitative testing EA-4/16, rev00 (Dec. 2003).
95 Ewan, W. D.; Kemp, K. W.: Sampling inspection of continuous processes with no autocorrelation between successive results, *Biometrika* 47 (1960), 363–380.
96 Ewan, W. D.: When and how to use Cusum Charts, *Technometrics* 5 (1963), 1–22.
97 Faltblatt: Trinkwasser-Ringversuche in Deutschland, *www.uni-stuttgart.de/siwa/ ch/aqs/pdf/TWFlyer.pdf* (September 2003).
98 Feinberg, M.: Basics of Interlaboratory Studies: the Trends in the New ISO 5725 Standard Edition, *Trends in Analytical Chemistry* 14 (1995), 450–457.
99 Franke, J. P.; de Zeeuw, R. A.: Evaluation and Optimization of the Standard Addition Method for Absorption Spectrometry and Anionic Stripping Voltammetry, *Anal. Chem.* 50 (1978), 1374–1380.

100 Funk, W.; Dammann, V.; Couturier, T.; Schiller, J., Völker, L.: Quantitative HPTLC Determination of Selenium, *HRC+CC* 9 (1986), 224–235.
101 Funk, W.; Dammann, V.; Vonderheid, C.; Oehlmann, G.: Statistische Methoden in der Wasseranalytik, VCH, Weinheim (1985).
102 Garfield, F. M.: Quality Assurance Principles for Analytical Laboratories, AOAC (1991), ISBN 0–935584-46–3.
103 Gesetz über Abgaben für das Einleiten von Abwasser in Gewässer (Abwasserabgabengesetz – AbwAG), Fassung vom 3. November 1994 (BGBl. I p. 3370; 1996 p. 1690; 1997 p. 582; 1998 p. 2455, 2001 p. 2331).
104 Gesetz zur Ordnung des Wasserhaushalts (WHG – Wasserhaushaltsgesetz) vom 19. August 2002 (BGBl. I Nr. 59 vom 23.8.2002 p. 3245; 6.1.2004 p. 2).
105 Glenn, G. C.; Hathaway, T. K.: Quality Control by Blind Sample Analysis, *Am. J. Clin. Path.* Vol. 72, No. 2 (1979), 156.
106 Goldsmith, P. L.; Whitfield, H.: Average Run Lengths in Cumulative Chart Quality Control Schemes, *Technometrics* 3 (1961), 11–21.
107 Gottschalk, G.: Einführung in die Grundlagen der chemischen Materialprüfung, Hirzel Verlag, Stuttgart, 1966.
108 Haeckel, R.; Höpfel, P.: Fehlersuche bei photometrischen Messungen im medizinischen Laboratorium, *GIT, Labor-Medizin* 80 (2) (1980), 112–115.
109 Haeckel, R.: Qualitätssicherung im medizinischen Labor, Deutscher Ärzte-Verlag, Fach-Taschenbuch Nr. 12, Köln (1975).
110 Hahn, J.: Meßinstrumentarium des Abwasserabgabe- und Wasserhaushaltsgesetzes, *Korr. Abw.* (1982), 670–679.
111 Handbook for Calculation of Measurement Uncertainty in Environmental Laboratories, Nordtest Project 1589–02 (2003). Available from Eurofins A/S, Agern Allé 11, DK-2970 Horsholm, Denmark.
112 Harbach, D.; Diehl, H.; Timm, J.; Huntemann, D.: Vergleich von drei Probenaufbereitungsverfahren für die quantitative Bestimmung von Blei in Fruchtsäften mit der flammlosen Atomabsorptionsspektrometrie, *Fresenius Z. Anal. Chem.* 301 (1980), 215–219.
113 Hartung, J.: Statistik, R. Oldenburg Verlag, München (1982).
114 Huber, L.: Validation and Qualification in Analytical Laboratories, Interpharm Press, Buffalo Grove (1999).
115 Huber, W.: Nachweis von Ausreißern und Nichtlinearitäten bei der Auswertung von Eichreihen über eine Regressionsrechnung, *Fresenius Z. Anal. Chem.* 319 (1984), 379–383.
116 Hund, E.; Massart, D. L.; Smeyers-Verbeke, J.: Operational Definitions of Uncertainty, *Trends in Analytical Chemistry* 20 (2001), 394–406.
117 Inhorn, St. L. (Ed.): Quality Assurance Practices for Health Laboratories, American Public Health Association (1978).
118 ISO 11095: Linear calibration using reference materials (Feb. 1996).
119 ISO 11843–1: Capability of Detection – Part 1: Terms and Definitions (1997).
120 ISO 11843–2: Capability of Detection – Part 2: Methodology in the Linear Calibration Case (2000).
121 ISO 17381: Water quality – Selection and application of ready-to-use test kit methods in water analysis (Dec. 2003).
122 ISO 3534–1 Statistics – Vocabulary and Symbols – Part 1: Probability and general statistical terms.
123 ISO 78–2: Chemistry – Layouts for standards – Part 2: Methods of chemical analysis (1999–03).
124 ISO 7870: Control Charts – General Guide and Introduction (Dec. 1993).
125 ISO 7873: Control Charts for Arithmetic Average with Warning Limits (Dec. 1993).
126 ISO Guide 35: Certification of Reference Materials – General and Statistical Principles (1989).
127 ISO Guide 38: General Requirements for the Acceptance of Testing Laboratories.
128 ISO Guide 45: Guidelines for the Presentation of Test Results.
129 ISO Guide 49: Guidelines for the Development of a Quality Manual for Testing Laboratories.
130 ISO Guide 30: Terms and Definitions used in Connection with Reference Materials.

131 ISO Guide 31: Contents of Certificates of Reference Materials (1981).
132 ISO/CD 11843–3: Capability of Detection – Part 3: Methodology for Determination of the Critical Value for the Response Variable when no Calibration Data are Used (2003).
133 ISO/CD 11843–4: Capability of Detection – Part 4: Methodology for Comparing the Minimum Detectable Value with a Given Value (2000).
134 ISO/TS 16489 (05/2006): Guidance for Establishing the Equivalency of Results.
135 ISO/TS 21748: Guidance for the Use of Repeatability, Reproducibility, and Trueness Estimates in Measurement Uncertainty Estimation (2004).
136 ISO/CD 13530: Water Quality – Guide to Analytical Quality Control for Water Analysis (2006).
137 IUPAC Technical Report (Draft): Harmonised Guidelines for the In-House Validation of Methods of Analysis. (http://www.iupac.org/divisions/V/501/draftoct19.pdf, 01/17/02).
138 Jackwerth, E.: Zur Eliminierung systematischer Fehler: Möglichkeiten und Probleme des Standard-Additionsverfahrens, *Chemie für Labor und Betrieb* 33 (1982), 4–10.
139 Jardine, A. K. S.; MacFarlane, J. D.; Greensted, C. S.: Statistical Methods for Quality Control, The Pitman Press, Bath, U.K. (1975).
140 Jork, H.; Funk, W.; Fischer, W.; Wimmer, H.: Thin-Layer Chromatography, Vol. 1A, Physical and Chemical Detection Methods, Wiley-VCH, Weinheim (February 1998).
141 Juran, J. D.; Gryna, F. M.; Bingham, R. S. (Eds.): Quality Control Handbook, McGraw-Hill Book Company, New York (1975).
142 Kateman, G.; Buydens, M.; Buydens, L.: Quality Control in Analytical Chemistry, Wiley Interscience (1993).
143 Kemp, K. W.: The Use of Cumulative Sums for Sampling Inspection Schemes, *Applied Statistics* 11 (1962), 16–31.
144 Kenkel, J.: A Primer on Quality in the Analytical Laboratory, Lewis Publ., Boca Raton, London, New York, Washington (2000).

145 Klärschlammverordnung (AbfKlärV) vom 15. April 1992 (BGBl. I 1992 p. 912; 1997 p. 446; 25. 3. 2002 p. 1193; 25. 4. 2002 p. 1488; 26. 11. 2003 p. 2373).
146 Krutz, H.; Cammann, K.; Donnevert, G.; Funk, W.; Hebbel, H.; Kolloch, B.; Laubereau, P. G.; Leichtfuß, S.; Neitzel, V.; Rump, H. H.: Chemometrie in der Wasseranalytik, *Vom Wasser* 72 (1989), 125–143.
147 Lanser, T. R.; Ballinger, D. G.: (EPA, USA) Quality Assurance Update, *Environmental Science & Technology* 13 (1979), 1356–1366.
148 Lucas, J. M.: A Modified „V-Mask" Control Scheme, *Technometrics* 15 (1973), 833–847.
149 Mager, H.: Moderne Regressionsanalyse, Salle & Sauerländer, Frankfurt (1982).
150 Mandel, J.; Lashof, T. W.: Interpretation and Generalization of Youden's Two-Sample Diagram, *Journal of Quality Technology*, Vol. 6, No. 1 (Jan. 1974), 22–36.
151 Mandel, J.: The Statistical Analysis of Experimental Data, Interscience Publ., J. Wiley & Sons, New York (1964).
152 Markowetz, D.: Qualitätsmanagement in der Laboratoriumsmedizin, Chapman & Hall, Weinheim (1997).
153 Massart, D. L.; Vandeginste, B. G. M.; Buydens, L. M. C.; De Jong, S.; Lewi, P. J.; Smeyers-Verbeke, J.: Handbook of Chemometrics and Qualimetrics, Parts A and B, Elsevier Science, Amsterdam (1997).
154 Massart, D. L.: Evaluation and Optimization of Laboratory Methods and Analytical Procedures, Elsevier Scientific Publishing Co., Amsterdam (1978).
155 Meier, P. C.; Zünd, R. E.: Statistical Methods in Analytical Chemistry, J. Wiley, New York, 2nd Edition (2000).
156 Merkblatt zu den Rahmenempfehlungen der Länderarbeitsgemeinschaft Wasser (LAWA) für die Qualitätssicherung bei Wasser-, Abwasser-, und Schlammuntersuchungen Nr. A-2 Kontrollkarten (Gelbdruck, January 2004).
157 Merkblatt zu den Rahmenempfehlungen der Länderarbeitsgemeinschaft Wasser (LAWA) für die Qualitätssicherung bei Wasser-, Abwasser-, und Schlammuntersuchungen Nr. A-3 (Stand Jan. 2001):

Ringversuche zur externen Qualitätsprüfung von Laboratorien.
158 Merkblatt zur Rahmenempfehlung der Länderarbeitsgemeinschaft Wasser (LAWA) für die Qualitätssicherung bei der Wasser-, Abwasser-, und Schlammuntersuchung Nr. A-4 (Stand 1.3.1989): Plausibilitätskontrolle.
159 Miller, J. C.; Miller, J. N.: Statistics for Analytical Chemistry, Second edition, Ellis Horwood Limited, Chichester, England (1988).
160 Mitchell, D. G.; Mills, W. N., Garden, J. S.: Multiple-Curve Procedure for Improving Precision with Calibration-Curve-Based Analyses, *Anal. Chem.* 49 (1977), 1655–1660.
161 Morries, P.; Hunt, D. T. E.: The Determinant Content of Water used in Blank Determinations, Water Research Centre, Medmenham, U.K. (May 1982).
162 Müller, C. H.; Uhlig, S.: Estimation of Variance Components with High Breakdown Point and High Efficiency, *Biometrika* 88 (2001), 353–366.
163 Natrella, M. G.: Experimental Statistics, National Bureau of Standards Handbook 91, Washington D.C. (1966).
164 Naturvårdsverket Rapport 3377: Interkalibrering 1987–1, Jonbalans, konduktivitet och pH, Solna, Sweden (1987).
165 OECD Series on Principles of Good Laboratory Practice and Compliance Monitoring, No. 1: OECD Principles of Good Laboratory Practice, Environment Monograph No. 45, Paris (1998).
166 OECD Series on Principles of Good Laboratory Practice and Compliance Monitoring, No. 10: Application of the Principles of GLP to Computerized Systems, Paris (1995).
167 OECD Series on Principles of Good Laboratory Practice and Compliance Monitoring, No. 13: Consensus Document: The Application of the OECD Principles of GLP to the Organization and Management of Multi-Site Studies, Paris (2002).
168 OECD Series on Principles of Good Laboratory Practice and Compliance Monitoring, No. 2: Guidance for GLP Monitoring Authorities. Revised Guides for Compliance Monitoring Procedures for Good Laboratory Practice, Paris (1995).
169 OECD Series on Principles of Good Laboratory Practice and Compliance Monitoring, No. 3: Guidance for GLP Monitoring Authorities. Revised Guidance for the Conduct of Laboratory Inspections and Study Audits, Environment Monograph No. 111, Paris (1995).
170 OECD Series on Principles of Good Laboratory Practice and Compliance Monitoring, No. 4: Consensus Document: Quality Assurance and GLP, Paris (1999).
171 OECD Series on Principles of Good Laboratory Practice and Compliance Monitoring, No. 5 (revised): Consensus Document: Compliance of Laboratory Suppliers with GLP Principles, Paris (1999).
172 OECD Series on Principles of Good Laboratory Practice and Compliance Monitoring, No. 6 (revised): Consensus Document: The Application of the GLP Principles to Field Studies, Paris (1999).
173 OECD Series on Principles of Good Laboratory Practice and Compliance Monitoring, No. 7 (revised): Consensus Document: The Application of the GLP Principles to Short-Term Studies, Paris (1999).
174 OECD Series on Principles of Good Laboratory Practice and Compliance Monitoring, No. 8 (revised): Consensus Document: The Role and Responsibilities for the Study Director in GLP Studies, Paris (1999).
175 Page, E. S.: Continuous Inspection Schemes, *Biometrika* 41 (1954), 100–115.
176 Papke, G.: Beispiele für die Anwendung der analytischen Qualitätssicherung (AQS) bei der Bestimmung des CSB, 10th Aachener Werkstattgespräch, 25–26th September 1986; *gwa* 92 (1987), 58–81.
177 Prichard, E.: Quality in the Analytical Chemistry Laboratory, J. Wiley, Chichester (1995).
178 Ratzlaff, K. L.: Optimizing Precision in Standard Addition Measurement, *Anal. Chem.* 51 (1979), 232–235.
179 Reed, A. H.; Henry, R. J.: Accuracy, Precision, Quality Control, and Miscella-

neous Statistics, Clinical Chemistry Principles and Techniques, Chapter 12, 287–341.
180 Reeuwijk, L. P. van; Houba, V. J. G.: Guidelines for Quality Management in Soil and Plant Laboratories, FAO and ISRIC, Rome (1998).
181 Commission Directive 1999/11/EC of 8th March 1999 adapting to technical progress the principles of good laboratory practice as specified in Council Directive 87/18/EEC on the harmonisation of laws, regulations and administrative provisions relating to the application of the principles of good laboratory practice and the verification of their applications for tests on chemical substances.
182 Commission Directive 1999/12/EC of 8th March 1999 adapting to technical progress for the second time the Annex to Council Directive 88/320/EEC on the inspection and verification of good laboratory practice (GLP).
183 Commission Directive 2003/78/EC of 11th August 2003 laying down the sampling methods and the methods of analysis for the official control of the levels of patulin in foodstuffs.
184 Council Directive 88/320/EEC of 9th June 1988 on the inspection and verification of Good Laboratory Practice (GLP), *Official Journal* L 145, 11/06/1988, p. 0035–0037.
185 Commission Directive 98/53/EC of 16th July 1998 laying down the sampling methods and the methods of analysis for the official control of the levels for certain contaminants in foodstuffs, *Official Journal* L 201, 17/07/1998, p. 0093–0101.
186 Council Directive 98/83/EC of 3rd November 1998 on the quality of water intended for human consumption, *Official Journal* L 330, 05/12/1998, p. 0032–0054.
187 Richtlinie der Bundesärztekammer zur Qualitätssicherung quantitativer laboratoriumsmedizinischer Untersuchungen (ab 01.01.2002), *http://www.bundesaerztekammer.de*
188 Richtlinien der Bundesärztekammer zur Qualitätssicherung in der Mikrobiologie (Teil A und B) (1992).
189 Rules Governing Medicinal Products in the European Community, Volume IV: *Good Manufacturing Practice for Medicinal Products*; III/3093/92-EN, January 1992.
190 Sachs, L.: Angewandte Statistik: Anwendung statistischer Methoden, 9th revised edition, Springer Verlag, Berlin, Heidelberg, New York (1999).
191 Seydlitz, G. v.: II. Fehlermöglichkeiten und systematische Fehlersuche, Wissenschaftliche Abteilung der ASID Bonz and Sohn GmbH, 85716 Unterschleißheim.
192 Sharaf, M. A.; Illmann, D. L.; Kowalski, B. R: Chemometrics, John Wiley & Sons, New York (1986).
193 Shewhart, W.: The Economic Control of Quality of Manufactured Product, D. van Nostrand Company. Inc., New York (1931).
194 Sokolowski, G.; Wood, W. G.: Radioimmunoassay in Theorie und Praxis, Schnetztor-Verlag, Konstanz (1981).
195 Stamm, D.: Calibration and Quality Control Materials, *Z. Klin. Chem. Klin. Biochem.* 12 (1974), 137–145.
196 Strategien für die Wasseranalytik: Verfahrensentwicklung, Validierung und Qualitätssicherung in der Routine. *Deutsche Einheitsverfahren zur Wasser-, Abwasser-, und Schlammanalytik* (DEV), 39. Lieferung, Verlag Chemie, Weinheim (1997).
197 Thompson, M.; Ellison, S. L. R.; Wood, R.: Harmonized Guidelines for Single-Laboratory Validation of Methods of Analysis (IUPAC Technical Report), *Pure & Appl. Chem.* Vol. 74, No. 5 (2002), 835–855.
198 Thompson, M.; Wood, R.: The International Harmonized Protocol for the Proficiency Testing of (Chemical) Analytical Laboratories – Results from the Symposium on Harmonization of Quality Assurance Systems in Chemical Analysis, Geneva, Switzerland, May 1991 (IUPAC, ISO, AOAC); *Pure & Appl. Chem.* 65 (1993), 2123–2144.
199 Timm, J.; Diehl, H.; Harbach, D.: Voraussetzung und Grenzen zur Anwendung der Additionsmethode bei der Atomabsorptions-Spektrometrie, *Fresenius Z. Anal. Chem.* 301 (1980), 199–202.

200 Tonks, D. B.: A Dual Program of Quality Control for Clinical Chemistry Laboratories, with a Discussion of Allowable Limits of Error; *Z. Anal. Chem.* 243 (1968), 760.

201 Uhlig, S.: Robust Estimation of Variance Components with High Breakdown Point in the 1-Way Random Effect Model, in: Industrial Statistics (Eds.: Kitsos, C. P.; Edler, L.), Physica-Verlag, p. 65–73 (1997).

202 van Dobben de Bruyn, C. S.: Cumulative Sum Techniques: Theory and Practice, Griffin's Statistical Monographs and Courses, No. 24, Charles Griffin, London (1968).

203 VDI 2449 Blatt 3: Prüfkriterien von Messverfahren, Allgemeine Methode zur Ermittlung der Unsicherheit kalibrierfähiger Messverfahren (2001).

204 Verordnung über Anforderungen an das Einleiten von Abwasser in Gewässer AbwV – Abwasserverordnung – vom 15. Oktober 2002 (BGBl. Nr. I vom 23.10.2002 S. 4047, ber. 2002 S. 4550)

205 Verordnung über die Qualität von Wasser für den menschlichen Gebrauch (TrinkwV 2001 – Trinkwasserverordnung) vom 21. Mai 2001, BGBl. I Nr. 24 vom 28.5.2001 p. 959; 25.11.2003 p. 2304.

206 Wassergesetz für das Land Nordrhein-Westfalen (Landeswassergesetz – LWG) i. d. F. d.Bek. v. 9. Juni 1989 (GV.NW.S. 384).

207 Wenclawiak, B. W.; , Koch, M.; Hadjicostas, E.: Quality Assurance in Analytical Chemistry – Training and Teaching, Springer, Berlin, Heidelberg, New York (2004).

208 Wernimont, G.: Use of Control Charts in the Analytical Laboratory, *Industrial and Engineering Chemistry, Analytical Edition* 18 (1946), 587–592.

209 Westgard, J. O.; Hunt, M. R.: Use and Interpretation of Common Statistical Tests in Method – Comparison Studies. *Clin. Chem.* 19 (1973), 49–57.

210 Wetherill, G. B.: Sampling Inspection and Quality Control, Chapman and Hall, London, New York, 2nd edition (1982).

211 Whitehead, T. P., et al.: Quality Control in Clinical Chemistry, John Wiley & Sons, New York, London (1977).

212 WHO: EuroReports "Quality Assessment in Health Laboratories" (1981).

213 WHO Good Manufacturing Practice for Pharmaceutical Products; PHARM/ 90.129 Rev. 3, March 1991.

214 Wilrich, P.-Th.: Qualitätsregelkarte bei vorgegebenen Grenzwerten, *QZ* 24, 10 (1979), 260–280.

215 Wilson, A. L.: The Performance Characteristics of Analytical Methods – II, *Talanta* 17 (1970), 30–44.

216 Woodward, R. H.; Goldsmith, P. L.: Cumulative Sum Techniques, Mathematical and Statistical Techniques for Industry, Monograph No. 3, Oliver and Boyd for ICI, Edinburgh (1972).

217 Youden, W. J.; Steiner, E. H.: Statistical Manual of the Association of Official Analytical Chemists, AOAC Publication, 2nd Edition (1979).

218 Youden, W. J.: Graphical Diagnosis of Interlaboratory Test Results, *Journal of Quality Technology,* Vol. 4, No. 1 (January 1972), 29–33.

219 Zoonen, P. van; Hoogerbrugge, R.; Gort, s. M.; Wiele, H. J. van de: Some Practical Examples of Method Validation in the Analytical Laboratory, *Trends in Analytical Chemistry* 18 (1999), 584–593.

Appendix 1

A1
Sample Calculations

The following sample calculations include examples of all *statistical* internal quality assurance procedures. A fictitious analytical process, called the XYZ process, is purposely used to illustrate as many problem situations as possible (too high an imprecision, systematic deviations of all sorts, serious matrix influences, etc.).

Comment on "computer notation": Very large or very small values are represented in floating point notation with the exponent of 10 given.

Example:

0.42193E-5 $\triangleq$ 0.42193 · 10^{-5} (= 0.0000042193)
6.21539E4 $\triangleq$ 6.21539 · 10^{4} (= 62153.9)

A1.1
Fundamental Calibration

Since the XYZ process is used to analyze waste water with substance contents of around 170 mg/l, an initial range of 100 to 280 mg/l is considered reasonable. Ten evenly spaced standard concentrations (x_1 to x_{10}) are prepared, treated with reagent, and analyzed photometrically.
The two extreme standards (x_1 and x_{10}) are each analyzed ten times.
The first- and second-order calibration functions are calculated; see Table A.1.1.

Linear regression:

$$Q_{xx} = \sum x_i^2 - \frac{1}{N} \cdot \left(\sum x_i\right)^2 \qquad = 33000 \ (\text{mg/l})^2$$

$$Q_{yy} = \sum y_i^2 - \frac{1}{N} \cdot \left(\sum y_i\right)^2 \qquad = 0.9829 \ \text{abs.}^2$$

$$Q_{xy} = \sum (x_i \cdot y_i) - \left(\frac{1}{N} \cdot \sum x_i \cdot \sum y_i\right) = 179.9400 \ \text{abs.} \cdot (\text{mg/l})$$

Quality Assurance in Analytical Chemistry: Applications in Environmental, Food, and Materials Analysis, Biotechnology, and Medical Engineering, Second Edition. W. Funk, V. Dammann, G. Donnevert
Copyright © 2007 WILEY-VCH Verlag GmbH & Co. KGaA, Weinheim
ISBN: 978-3-527-31114-9

Table A1.1 Measured data and calculation table (for reasons of clarity, the exponential notation E has been substituted for large individual values).

No.	x values	y values	x^2	x^3	x^4	y^2	$x \cdot y$	$x^2 \cdot y$
1	100	1.064	1.000000E4	1.000000E6	1.000000E8	1.132096E0	1.064000E2	1.064000E4
2	120	1.177	1.440000E4	1.728000E6	2.073600E8	1.385329E0	1.412400E2	1.694880E4
3	140	1.303	1.960000E4	2.744000E6	3.841600E8	1.697809E0	1.824200E2	2.553880E4
4	160	1.414	2.560000E4	4.096000E6	6.553600E8	1.999396E0	2.262400E2	3.619840E4
5	180	1.534	3.240000E4	5.832000E6	1.049760E9	2.353156E0	2.761200E2	4.970160E4
6	200	1.642	4.000000E4	8.000000E6	1.600000E9	2.696164E0	3.284000E2	6.568000E4
7	220	1.744	4.840000E4	1.064800E7	2.342560E9	3.041536E0	3.836800E2	8.440960E4
8	240	1.852	5.760000E4	1.382400E7	3.317760E9	3.429904E0	4.444800E2	1.066752E5
9	260	1.936	6.760000E4	1.757600E7	4.569760E9	3.748096E0	5.033600E2	1.308736E5
10 = N	280	2.046	7.840000E4	2.195200E7	6.146560E9	4.186116E0	5.728800E2	1.604064E5
Sum:	1900.00	15.712	3.940000E5	8.740000E7	2.037328E10	2.566960E1	3.165220E3	6.870724E5

Mean $\bar{x}$:	$\bar{x} = \dfrac{1}{N}\sum x_i$	$= 190$ mg/l
Mean $\bar{y}$:	$\bar{y} = \dfrac{1}{N}\sum y_i$	$= 1.5712$ abs.
Slope:	$b = \dfrac{Q_{xy}}{Q_{xx}}$	$= 0.00545\ \dfrac{\text{abs.}}{\text{mg/l}}$
Axis intercept:	$a = \bar{y} - b\cdot\bar{x}$	$= 0.5352$ abs.
Residual standard deviation:	$s_y = \sqrt{\dfrac{1}{N-2}\cdot\left(Q_{yy} - \dfrac{Q_{x^2y}}{Q_{xx}}\right)}$	$= 0.01476$ abs.
Process standard deviation:	$s_{xo} = \dfrac{s_y}{b}$	$= 2.708$ mg/l
Process variation coefficient:	$V_{xo} = \dfrac{s_{xo}}{\bar{x}}\cdot 100\%$	$= 1.43\%$
Test value x_a:	$y_a = a + s_y\cdot t\sqrt{\dfrac{1}{N} + 1 + \dfrac{\bar{x}^2}{Q_{xx}}}$	$= 0.5759$ abs.
	$x_a = 2\cdot\dfrac{y_a - a}{b}$	$= 14.92$ mg/l

where $t\ (f = N-2 = 8,\ P = 95\%,\ \text{one-sided}) = 1.860$

A1.2
Linearity Tests

A1.2.1 Visual Linearity Test

A graphical representation of the calibration data indicates nonlinearity.

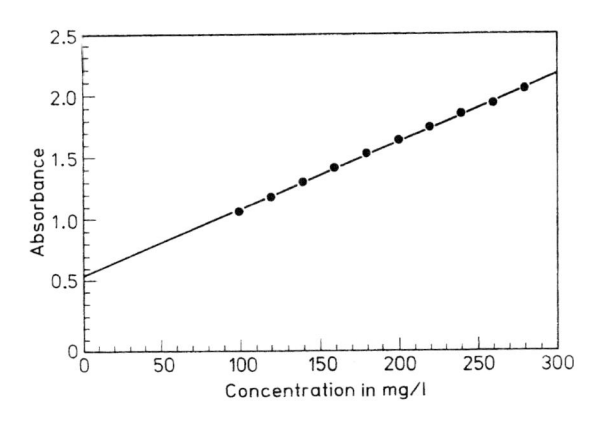

Fig. A1 First calibration curve.

Appendix 1

The graphical representation of the residuals indicates a systematically curved progression.

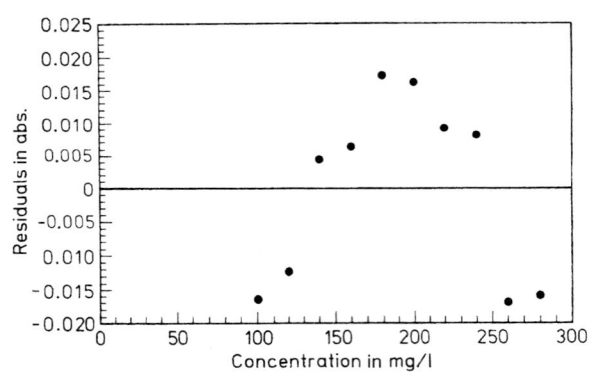

Fig. A2 Residuals of the first calibration function.

The second-order calibration function (curved) is calculated from the calibration data (Table A1.1).

A1.2.2 Second-Order Calibration Function

Second-order regression:

$$Q_{xx} = \sum x_i^2 - \frac{1}{N} \cdot \left(\sum x_i\right)^2 \qquad = 33000 \text{ (mg/l)}^2$$

$$Q_{xy} = \sum (x_i \cdot y_i) - \left(\frac{1}{N} \cdot \sum x_i \cdot \sum y_i\right) \qquad = 179.9400 \text{ abs.} \cdot \text{(mg/l)}$$

$$Q_{x^3} = \sum x_i^3 - \left(\frac{1}{N} \cdot \sum x_i \cdot \sum x_i^2\right) \qquad = 1.254000\text{E}7 \text{ (mg/l)}^3$$

$$Q_{x^4} = \sum x_i^4 - \left(\frac{1}{N} \left(\sum x_i^2\right)^2\right) \qquad = 4.849680\text{E}9 \text{ (mg/l)}^4$$

$$Q_{x^2y} = \sum (x_i^2 \cdot y_i) - \left(\frac{1}{N} \cdot \sum y_i \cdot \sum x_i^2\right) = 6.80196\text{E}4 \text{ (mg/l)}^2 \cdot \text{abs.}$$

$$\bar{x} = \frac{1}{N} \cdot \sum x_i \qquad = 190 \text{ mg/l}$$

$$\bar{y} = \frac{1}{N} \cdot \sum y_i \qquad = 1.5712 \text{ abs.}$$

Coefficients of the regression function:

$$c = \frac{Q_{xy} \cdot Q_{x^3} - Q_{x^2y} \cdot Q_{xx}}{(Q_{x^3})^2 - Q_{xx} \cdot Q_{x^4}} \qquad = -0.42329\text{E-}5 \; \frac{\text{abs.}}{(\text{mg/l})^2}$$

$$b = \frac{Q_{xy} - c \cdot Q_{x^3}}{Q_{xx}} \qquad = 0.007061 \; \frac{\text{abs.}}{(\text{mg/l})}$$

$$a = \bar{y} - b \cdot \bar{x} - \frac{c}{N} \sum x_i^2 \qquad = 0.39634 \text{ abs.}$$

Second-order calibration function:

$$y = 0.39634 \text{ abs.} + 0.007061 \frac{\text{abs.}}{(\text{mg/l})} \cdot x - 0.42329\text{E-}5 \frac{\text{abs.}}{(\text{mg/l})^2} \cdot x^2$$

Residual standard deviation:

$$s_y = \sqrt{\frac{1}{N-3} \cdot \left(\sum y_i^2 - a \cdot \sum y_i - b \cdot \sum x_i y_i - c \cdot \sum x_i^2 y_i \right)} = 0.005734 \text{ abs.}$$

Note: If the calibration data in question and the above calculated second-order calibration function are to be seen as definitive, then additional data should be calculated:

Sensitivity: $\qquad E = b + 2 \cdot c \cdot \bar{x} = 0.005453 \frac{\text{abs.}}{(\text{mg/l})}$

Process standard deviation: $s_{xo} = \frac{s_y}{E} = 1.0516 \text{ mg/l}$

Analytical Results for the Second-Order Calibration Function

For a measured value of $\hat{y} = 1.100$ [abs.], an analytical result

$$\hat{x} = -\frac{b}{2c} - \sqrt{\left(\frac{b}{2c}\right)^2 - \frac{a - \hat{y}}{c}} = 106.443 \text{ mg/l}$$

would be obtained because $c < 0$ (negative curvature).

$$\left[\begin{array}{l} \text{For positive } c \text{ (positive curvature), } \hat{x} \text{ would become} \\[6pt] \hat{x} = -\dfrac{b}{2c} + \sqrt{\left(\dfrac{b}{2c}\right)^2 - \dfrac{a - \hat{y}}{c}} = 94.321 \text{ mg/l} \\[6pt] \text{or, in general: } \hat{x} = -\dfrac{b}{2c} + \left[signum(c) \cdot \sqrt{\left(\dfrac{b}{2c}\right)^2 - \dfrac{a - \hat{y}}{c}} \right] \end{array} \right]$$

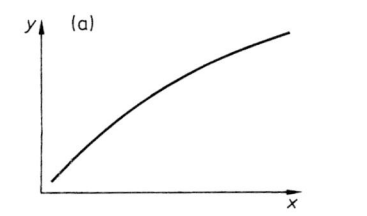

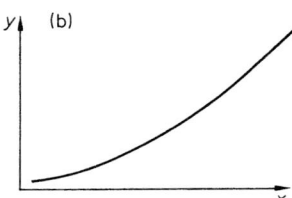

Fig. A3 Curvature of nonlinear calibration curves:
a) negative curvature, b) positive curvature.

Given t $(f = N_c - 3, P = 95\%) = 2.364$, then the confidence interval amounts to

$$CI(\hat{x}) = \frac{s_y \cdot t}{(b + 2c\hat{x})} \cdot \sqrt{\frac{1}{N_c} + \frac{1}{N_a} + \frac{1}{Q_{x^4} \cdot Q_{xx} - (Q_{x^3})^2}}$$

$$\cdot \left\{ (\hat{x} - \bar{x})^2 Q_{x^4} + \left(\hat{x}^2 - \frac{\sum x_i^2}{N_c} \right)^2 Q_{xx} - 2 \cdot (\hat{x} - \bar{x}) \cdot \left(\hat{x}^2 - \frac{\sum x_i^2}{N_c} \right) \cdot Q_{x^3} \right\}$$

$$= 2.670 \text{ mg/l}$$

and the final result is therefore

$$\hat{x} \pm CI(\hat{x}) = 106.447 \text{ mg/l} \pm 2.670 \text{ mg/l}$$

or $\quad \hat{x} \pm \dfrac{CI(\hat{x})}{\hat{x}} \cdot 100\% = 106.447 \text{ mg/l} \pm 2.51\%$

A1.2.3 Linearity Test: Goodness-of-Fit Test

The residual standard deviation of the first-order calibration function is:

$s_{y1} = 0.01476$ abs.

The residual standard deviation of the second-order calibration function is:

$s_{y2} = 0.005734$ abs.
$N = 10$
$DS^2 = (N - 2) \cdot s_{y1}^2 - (N - 3) \cdot s_{y2}^2 = 0.00151$ abs.

F-test:

$$TV = \frac{DS^2}{s_{y2}^2} = 46.0$$

$F(f_1 = 1, f_2 = 7, P = 99\%) = 12.25$

Decision: Since $TV \gg F$, the second-order regression achieves a significantly better fit.

As a consequence of the test results, all analytical processes are systematically tested: a correction of the amounts of reagent added finally results in a linear calibration function:

A1 Sample Calculations

Table A1.2 Measured data (after correction) and calculation table.

No.	x values	y values	x^2	x^3	x^4	y^2	$x \cdot y$	$x^2 \cdot y$
1	100	0.752	1.000000E4	1.000000E6	1.000000E8	5.655040E-1	7.520000E1	7.520000E3
2	120	0.890	1.444000E4	1.728000E6	2.073600E8	7.921000E-1	1.068000E2	1.281600E4
3	140	1.037	1.960000E4	2.744000E6	3.841600E8	1.075369E0	1.451800E2	2.032520E4
4	160	1.168	2.560000E4	4.096000E6	6.553600E8	1.364224E0	1.868800E2	2.990080E4
5	180	1.324	3.240000E4	5.832000E6	1.049760E9	1.752976E0	2.383200E2	4.289760E4
6	200	1.447	4.000000E4	8.000000E6	1.600000E9	2.093809E0	2.894000E2	5.788000E4
7	220	1.604	4.840000E4	1.064800E7	2.342560E9	2.572816E0	3.528800E2	7.763360E4
8	240	1.726	5.760000E4	1.382400E7	2.979076E9	2.979076E0	4.142400E2	9.941760E4
9	260	1.885	6.760000E4	1.757600E7	3.553225E9	3.553225E0	4.901000E2	1.274260E5
10	280	2.001	7.840000E4	2.195200E7	6.146560E9	4.004001E0	5.602800E2	1.568784E5
Sum:	1900.00	13.834	3.940000E5	8.740000E7	2.037328E10	2.075310E1	2.859280E3	6.326952E5

Linear regression:

Q_{xx}:	33000.0000 (mg/l)^2
Q_{yy}:	1.6151 abs.^{2}
Q_{xy}:	230.8200 abs. (mg/l)
mean $\bar{x}$:	190.0000 mg/l
mean $\bar{y}$:	1.3834 abs.
slope b:	0.0070 abs./(mg/l)
axis intercept a:	0.0544 abs.
residual standard deviation s_y:	0.009 abs.
process standard deviation s_{xo}:	1.302 mg/l
process variation coeff. V_{xo}:	0.69 %

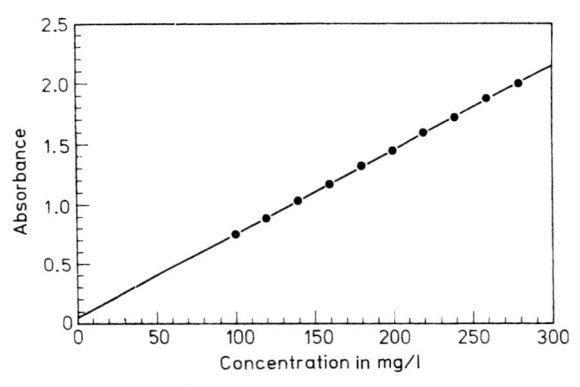

Fig. A4 Second calibration curve.

A1.2.4 Variance Homogeneity Test

The graphical representation of calibration data and residuals suggests that the variance homogeneity should be verified.

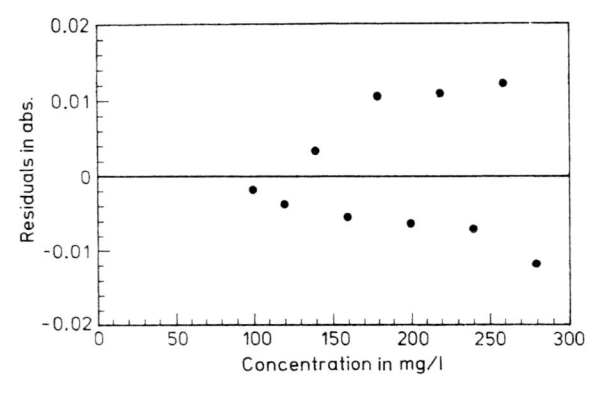

Fig. A5 Residuals of the second calibration function.

The two standards, $x_1 = 100$ mg/l and $x_{10} = 280$ mg/l, are each analyzed ten times.

y_1	y_{10}
0.538	2.005
0.536	1.992
0.537	2.007
0.539	1.994
0.534	1.995
0.535	2.003
0.534	1.990
0.537	2.002
0.535	1.996
0.533	2.005

The means, $\bar{y}_1 = 0.5358$ abs., $\bar{y}_{10} = 1.9989$ abs., and the standard deviations, $s_1 = 0.00193$ abs., $s_{10} = 0.00615$ abs., are calculated for each measurement set and are submitted to an F-test:

$$TV = \frac{s_{10}^2}{s_1^2} = \frac{37.88 \cdot 10^{-6}}{3.733 \cdot 10^{-6}} = 10.15$$

$$F(f_1 = 9, f_2 = 9, P = 99\%) = 5.35$$

Decision: Since $TV > F$, one may assume variance inhomogeneity.

After reducing the range by nearly one-third (so that standard concentrations were only between 100 and 235 mg/l), new standard concentrations were prepared and the entire calibration was repeated.

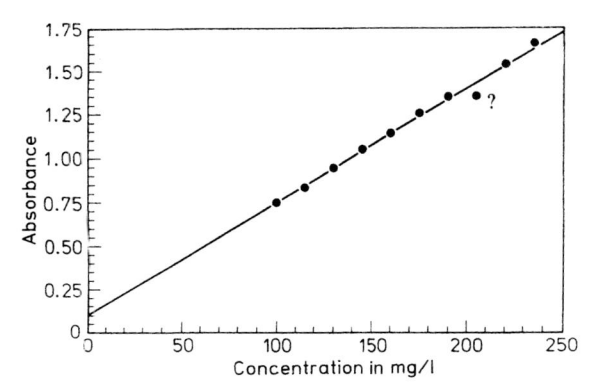

Fig. A6 Linear calibration function with one outlier.

A1.2.5 Outlier Tests for Linear Calibration

Calibration data:

x	y
100	0.754
115	0.842
130	0.950
145	1.063
160	1.148
175	1.264
190	1.352
205	1.360* suspected outlier
220	1.546
235	1.661

axis intercept: $a = 0.1028$ abs.
slope: $b = 0.006515$ abs./(mg/l)
residual standard deviation: $s_{y_{A1}} = 0.0314$ abs.

The linear regression calculation omitting the eighth measured value, a suspected outlier (marked with * in the table), yielded:

$a = 0.0801$ abs.
$b = 0.00671$ abs./(mg/l)
$s_{y_{A2}} = 0.00813$ abs.
$N_{A2} = 9$
$f_{A2} = 7$

To use the F-test as an outlier test, the test value is calculated:

$$TV = \frac{f_{A1} \cdot s^2_{y_{A1}} - f_{A2} \cdot s_{y_{A2}}}{s_{y_{A2}}} = \frac{0.007888 - 0.004627}{0.0000661} = 112.5$$

$F(f_1 = 1, f_2 = f_{A2} = 7, P = 99\%) = 12.25$

Since $TV > F$, the eighth standard concentration may be regarded as an outlier.

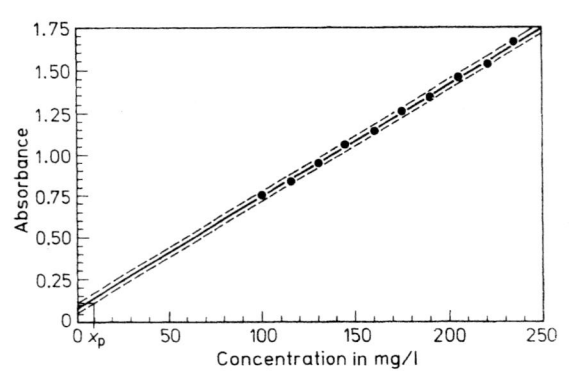

Fig. A7 Outlier t-test.

The *t*-test yielded the same results:

$a = 0.0801$ abs.
$b = 0.00671$ abs./(mg/l)
$s_{y_{A2}} = 0.00813$ abs.
$N_{A2} = 9$
$f_{A2} = 7$
$x_A = 205$ mg/l
$\hat{y}_A = a + b \cdot x_A = 1.455$ abs.
$Q_{xx} = 17000$ (mg/l)^2
$\bar{x} = 163.33$ mg/l

$$CI\left(\hat{y}_A\right) = \hat{y}_A \pm s_{y_{A2}} \cdot t \cdot \sqrt{1 + \frac{1}{N_{A2}} + \frac{(x_A - \bar{x})^2}{Q_{xx}}} = 1.455 \pm 0.021 \text{ abs.}$$

(where t ($f = 7$, $P = 95\%$) = 2.365)

Therefore, the suspected outlier value y_A = 1.360 abs. lies outside of the prognosis range. Because of the proven outlier, the entire calibration must be repeated and evaluated.

No.	x value	y value	x^2	y^2	$x \cdot y$
1	100	0.753	1.00000E4	5.67009E-1	7.53000E1
2	115	0.843	1.32250E4	7.10649E-1	9.69450E1
3	130	0.951	1.69000E4	9.04401E-1	1.23630E2
4	145	1.062	2.10250E4	1.12784E0	1.53990E2
5	160	1.149	2.56000E4	1.32020E0	1.83840E2
6	175	1.263	3.06250E4	1.59517E0	2.21025E2
7	190	1.353	3.61000E4	1.83061E0	2.57070E2
8	205	1.460	4.20250E4	2.13160E0	2.99300E2
9	220	1.545	4.84000E4	2.38702E0	3.39900E2
10	235	1.662	5.52250E4	2.76224E0	3.90570E2
Σ	1635	12.041	2.99125E5	1.53368E1	2.14157E3

The final characteristic data are obtained from these values:

Linear regression:

range:	100 to 235 mg/l
Q_{xx}:	18562.5000 (mg/l)^2
Q_{yy}:	0.8382 (abs.)^2
Q_{xy}:	124.7025 (mg/l) · abs.
mean $\bar{x}$:	167.5000 mg/l
mean $\bar{y}$:	1.2041 abs.
slope b:	0.0067 abs./(mg/l)
axis intercept a:	0.0788 abs.
residual standard deviation s_y:	0.0074 abs.

process standard deviation s_{xo}: 1.096 mg/l
process variation coefficient V_{xo}: 0.655%
$t\,(f=8,\,P=95\%,\text{ one-sided})$: 1.860
$t\,(f=8,\,P=95\%,\text{ two-sided})$: 2.306

A1.2.6 Securing the Lower Range Limit

The graph (Figure A8) already shows a test value, x_a, which lies markedly below x_1.

$(t\,(f=8,\,P=95\%) = 1.860)$

Mathematically, one likewise obtains the auxiliary value:

$$y_a = a + s_y \cdot t \sqrt{\frac{1}{N} + 1 + \frac{\bar{x}^2}{Q_{xx}}} = 0.10097 \text{ abs.}$$

as well as the test value

$$x_a = 2 \cdot \frac{y_a - a}{b} = 6.6 \text{ mg/l}$$

$x_a\,(=6.6 \text{ mg/l}) < x_1\,(=100 \text{ mg/l})$

Note: x_a corresponds to the "minimum detectable value", x_{MDV}, by calculation.

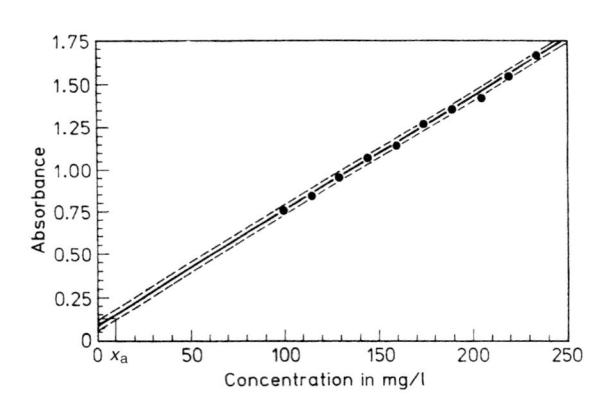

Fig. A8 Final linear calibration curve.

Given $t\,(f=8,\,P=95\%,\text{ two-sided}) = 2.306$, the relative analytical imprecision at the lower end of the range is

$$CI(x_1) = s_{xo} \cdot t \cdot \sqrt{\frac{1}{N} + 1 + \frac{(x_1 - \bar{x})^2}{Q_{xx}}} = 2.932 \text{ mg/l}$$

$$CI_{rel}(x_1) = \frac{CI(x_1)}{x_1} \cdot 100\% = 2.93\%$$

To illustrate this, both $CI(x)$ and $CI_{rel}(x)$ are plotted (see Figure A9). As an exception, $CI(x)$ is also calculated for x-values below x_1 (extrapolation).

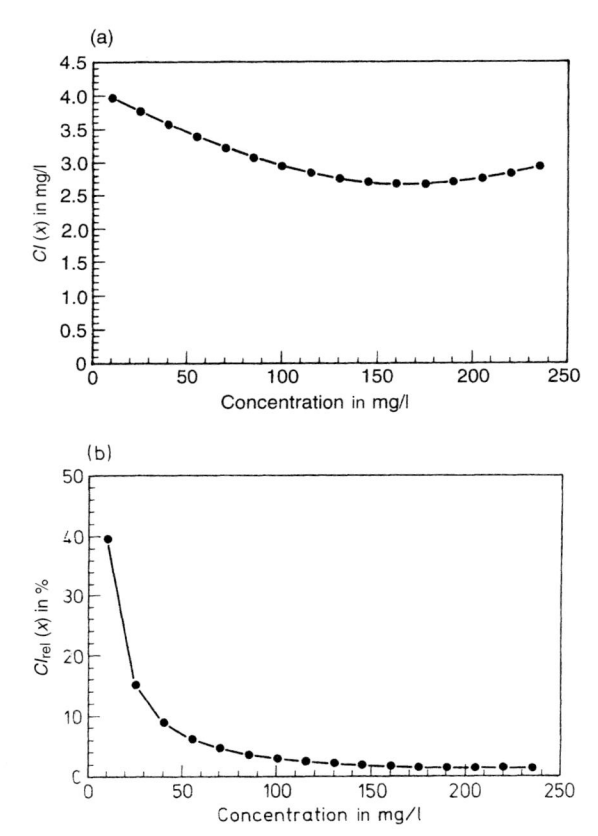

Fig. A9 Absolute and relative imprecision of the calibration:
a) course of $Cl(x)$ (also extrapolated),
b) course of $Cl_{rel}(x)$ (also extrapolated).

A1.2.7 Decision Limit, Minimum Detectable Value, and Limit of Quantification

The decision limit, minimum detectable value, and the limit of quantification are all obtained using either the distribution of blank values or the calibration function.

a) Blank value procedure

A blank sample was analyzed ten times and yielded the following results:

Measured value (absorbance):			
0.005	mean blank value:	$\bar{y}_B$	$= 0.0061$ abs.
0.007	standard deviation:	s_B	$= 0.00247$ abs.
0.006		N_B	$= 10$
0.009		N_a	$= 1$
		b	$= 0.0067$ abs./(mg/l)

0.007
0.008
0.003
0.001
0.008
0.007

$t\ (f = 9,\ P = 95\%,\ \text{one-sided}) = 1.833\ (t_\alpha = t_\beta)$

prediction interval: $\Delta y_B = s_B \cdot t_{f,\alpha} \sqrt{\dfrac{1}{N_a} + \dfrac{1}{N_B}}$
$= 0.00047\ \text{abs.}$

critical value: $y_c = \bar{y}_B + \Delta y_B = 0.0108\ \text{abs.}$

decision limit: $x_{DL} = \dfrac{s_B}{b} \cdot t_{f,\alpha} \sqrt{\dfrac{1}{N_a} + \dfrac{1}{N_B}}$
$= 0.71\ \text{mg/l}$

minimum detectable value: $x_{MDV} = 2 \cdot x_{DL} = 1.42\ \text{mg/l}$

limit of quantification: $x_{LQ} = 2 \cdot k \cdot s_B / b$ (with $k = 3$)
$x_{LQ} = 2.21\ \text{mg/l}$

b) Calibration function procedure

Calibrations were performed for four successively lower ranges and the decision limit, minimum detectable value, and limit of quantification were calculated.

First range from 30 to 75 mg/l.

No.	x value	y value	x^2	y^2	$x \cdot y$
1	30	0.155	9.00000E2	2.40250E-2	4.65000E0
2	35	0.163	1.22500E3	2.65690E-2	5.70500E0
3	40	0.192	1.60000E3	3.68640E-2	7.68000E0
4	45	0.216	2.02500E3	4.66560E-2	9.72000E0
5	50	0.227	2.50000E3	5.15290E-2	1.13500E1
6	55	0.259	3.02500E3	6.70810E-2	1.42450E1
7	60	0.272	3.60000E3	7.39840E-2	1.63200E1
8	65	0.291	4.22500E3	8.46810E-2	1.89150E1
9	70	0.305	4.90000E3	9.30250E-2	2.13500E1
10	75	0.327	5.62500E3	1.06929E-1	2.45250E1
Σ	525	2.407	2.96250E4	6.11343E-1	1.34460E2

Linear regression:

Q_{xx}:	2062.5000 (mg/l)^2
Q_{yy}:	0.0320 (abs.)^2
Q_{xy}:	8.0925 abs.·mg/l
mean $\bar{x}$:	52.5000 mg/l
mean $\bar{y}$:	0.2407 abs.
slope b:	0.0039 abs./(mg/l)
axis intercept a:	0.0347 abs.
residual standard deviation s_y:	0.005 abs.
process standard deviation s_{xo}:	1.35 mg/l

process variation coefficient V_{xo}: 2.58%
$t\ (f = 8, P = 95\%,\ \text{one-sided})$: 1.860
$t\ (f = 8, P = 95\%,\ \text{two-sided})$: 2.306
critical value y_c: 0.05014 abs.
"decision limit" x_{DL}: 3.93 mg/l
"minimum detectable value" x_{MDV}: 7.86 mg/l
"limit of quantification" $x_{LQ}\ (k = 3)$: 12.8 mg/l

Second range from 14 to 50 mg/l.

No.	x value	y value	x^2	y^2	$x \cdot y$
1	14	0.110	1.96000E2	1.21000E-2	1.54000E0
2	18	0.136	3.24000E2	1.84960E-2	2.44800E0
3	22	0.151	4.84000E2	2.28010E-2	3.32200E0
4	26	0.188	6.76000E2	3.53440E-2	4.88800E0
5	30	0.203	9.00000E2	4.12090E-2	6.09000E0
6	34	0.236	1.15600E3	5.56960E-2	8.02400E0
7	38	0.267	1.44400E3	7.12890E-2	1.01460E1
8	42	0.283	1.76400E3	8.00890E-2	1.18860E1
9	46	0.309	2.11600E3	9.54810E-2	1.42140E1
10	50	0.351	2.50000E3	1.23201E-1	1.75500E1
Σ	320	2.234	1.15600E4	5.55706E-1	8.01080E1

Linear regression:

Q_{xx}: 1320.0000 (mg/l)^2
Q_{yy}: 0.0566 (abs.)^2
Q_{xy}: 8.6200 abs. · mg/l
mean $\bar{x}$: 32.0000 mg/l
mean $\bar{y}$: 0.2234 abs.
slope b: 0.0065 abs./(mg/l)
axis intercept a: 0.0144 abs.
residual standard deviation s_y: 0.007 abs.
process standard deviation s_{xo}: 0.997 mg/l
process variation coefficient V_{xo}: 3.12%
$t\ (f = 8, P = 95\%,\ \text{one-sided})$: 1.860
$t\ (f = 8, P = 95\%,\ \text{two-sided})$: 2.306
critical value y_c: 0.03101 abs.
"decision limit" x_{DL}: 2.54 mg/l
"minimum detectable value" x_{MDV}: 5.08 mg/l
"limit of quantification" $x_{LQ}\ (k = 3)$: 8.51 mg/l

Appendix 1

Third range from 10 to 23.5 mg/l.

No.	x value	y value	x^2	y^2	$x \cdot y$
1	10.0	0.095	1.00000E2	9.02500E-3	9.50000E-1
2	13.0	0.115	1.69000E2	1.32250E-3	1.49500E0
3	16.0	0.131	2.56000E2	1.71610E-2	2.09600E0
4	19.0	0.139	3.61000E2	1.93210E-2	2.64100E0
5	22.0	0.155	4.84000E2	2.40250E-2	3.41000E0
6	25.0	0.163	6.25000E2	2.65690E-2	4.07500E0
7	28.0	0.181	7.84000E2	3.27610E-2	5.06800E0
8	31.0	0.190	9.61000E2	3.61000E-2	5.89000E0
9	34.0	0.207	1.15600E3	4.28490E-2	7.03800E0
10	37.0	0.217	1.36900E3	4.70890E-2	8.02900E0
Σ	235.5	1.593	6.26500E3	2.68125E-1	4.06920E1

Linear regression:

Q_{xx}:	742.500 (mg/l)^2
Q_{yy}:	0.01436 abs.^{2}
Q_{xy}:	3.2565 abs.·mg/l
mean $\bar{x}$:	23.500 mg/l
mean $\bar{y}$:	0.1593 abs.
slope b:	0.0044 abs./(mg/l)
axis intercept a:	0.0562 abs.
residual standard deviation s_y:	0.0031 abs.
process standard deviation s_{xo}:	0.7099 mg/l
process variation coefficient V_{xo}:	3.02%
t ($f = 8$, $P = 95\%$, one-sided):	1.860
t ($f = 8$, $P = 95\%$, two-sided):	2.306
critical value y_c:	0.06409 abs.
"decision limit" x_{DL}	1.79 mg/l
"minimum detectable value" x_{MDV}:	3.58 mg/l
"limit of quantification" x_{LQ} ($k = 3$):	6.04 mg/l

Fourth range from 5 to 18.5 mg/l.

No.	x value	y value	x^2	y^2	$x \cdot y$
1	5.0	0.034	2.50000E1	1.15600E-3	1.70000E-1
2	6.5	0.056	4.22500E1	3.13600E-3	3.64000E-1
3	8.0	0.061	6.40000E1	3.72100E-3	4.88000E-1
4	9.5	0.064	9.02500E1	4.09600E-3	6.08000E-1
5	11.0	0.081	1.21000E2	6.56100E-3	8.91000E-1
6	12.5	0.093	1.56250E2	8.64900E-3	1.16250E0
7	14.0	0.106	1.96000E2	1.12360E-2	1.48400E0
8	15.5	0.112	2.40250E2	1.25440E-2	1.73600E0
9	17.0	0.117	2.89000E2	1.36890E-2	1.98900E0
10	18.5	0.125	3.42250E2	1.56250E-2	2.31250E0
Σ	117.5	0.849	1.56625E3	8.04130E-2	1.12050E1

Linear regression:

Q_{xx}:	185.6250 (mg/l)^2
Q_{yy}:	0.0083 (abs.)^2
Q_{xy}:	1.2292 abs. · mg/l
mean $\bar{x}$:	11.7500 mg/l
mean $\bar{y}$:	0.0849 abs.
slope b:	0.0066 abs./(mg/l)
axis intercept a:	0.0071 abs.
residual standard deviation s_y:	0.005 abs.
process standard deviation s_{xo}:	0.741 mg/l
process variation coefficient V_{xo}:	6.3047 %
$t\,(f=8,\ P=95\%,\ \text{one-sided})$:	1.860
$t\,(f=8,\ P=95\%,\ \text{two-sided})$:	2.306
critical value y_c:	0.01948 abs.
"decision limit" x_{DL}:	1.87 mg/l
"minimum detectable value" x_{MDV}:	3.74 mg/l
"limit of quantification" x_{LQ} ($k=3$):	5.82 mg/l

Summary of the limits calculated using the calibration slopes:

Range in mg/l	x_{DL} in mg/l	x_{MDV} in mg/l	x_{LQ} in mg/l
100–235	3.29	–	–
30–75	3.93	7.86	12.8
14–50	2.54	5.08	8.51
10–37	1.79	3.58	6.04
5–18.5	1.87	3.74	5.82

The conditions of DIN 32645 – that the ratio of the calculated decision limit to the highest calibration value does not exceed a factor of 10 – are met only by the lowest range (5 to 18.5 mg/l). Thus:

decision limit: $x_{DL} = 1.87$ mg/l
minimum detectable value: $x_{MDV} = 3.74$ mg/l
limit of quantification ($k = 3$): $x_{LQ} = 5.82$ mg/l

(In the examples, $N = 10$ analyses were performed. A lesser number would lead to a correspondingly larger range of confidence and to a higher decision limit.)

A1.2.8 Recovery Function

Sample digestion was introduced as an additional sample preparation step. The 10 standard concentrations of the calibration (from 100 to 235 mg/l) were analyzed incorporating the digestion step and were evaluated using the known calibration function

$$y = 0.0788 \text{ abs.} + 0.0067 \frac{\text{abs.}}{\text{mg/l}} \cdot x$$

solved for x: $\quad x = \dfrac{y - 0.0788 \text{ abs.}}{0.0067 \text{ abs.}/(\text{mg/l})}$

The process standard deviation of the calibration yielded:

$$s_{xo_c} = 1.096 \text{ mg/l}$$

Data list:

No.	$x_{calib.} = x$	$x_{found} = y$	x^2	y^2	$x \cdot y$
1	100	87	1.00000E4	7.56900E3	8.70000E3
2	115	100	1.32250E4	1.00000E4	1.15000E4
3	130	113	1.69000E4	1.27690E4	1.46900E4
4	145	125	2.10250E4	1.56250E4	1.81250E4
5	160	141	2.56000E4	1.98810E4	2.25600E4
6	175	152	3.06250E4	2.31040E4	2.66000E4
7	190	166	3.61000E4	2.75560E4	3.15400E4
8	205	181	4.20250E4	3.27610E4	3.71050E4
9	220	193	4.84000E4	3.72490E4	4.24600E4
10	235	209	5.52250E4	4.36810E4	4.91150E4
Σ	1675	1467	2.99125E5	2.30195E5	2.62395E5

Linear regression:

Q_{xx}:	18562.5 (mg/l)^2
Q_{yy}:	14986.1 (mg/l)^2
Q_{xy}:	16672.5 (mg/l)^2
mean $\bar{x}_c$:	167.5 mg/l
mean $\bar{x}_f$:	146.7 mg/l

Characteristic data of the recovery function:

slope: $\quad b_f = 0.8982$

axis intercept: $\quad a_f = -3.7455$ mg/l

residual standard deviation: $\quad s_{y_f} = 1.181$ mg/l

standard deviation of the slope: $\quad s_{b_f} = \dfrac{s_{y_f}}{\sqrt{Q_{xx}}} = 0.00867$

standard deviation of the intercept: $\quad s_{a_f} = s_{y_f} \cdot \sqrt{\dfrac{1}{N_f} + \dfrac{\bar{x}_c^2}{Q_{xx}}} = 1.499$ mg/l

Testing precision by means of the F-test:

$$TV = \left(\dfrac{s_{y_f}}{s_{xo_c}}\right)^2 = \left(\dfrac{1.181 \text{ mg/l}}{1.096 \text{ mg/l}}\right)^2 = 1.16$$

$F(f_1 = f_2 = 8, P = 99\%) = 6.03$

Since the test value of 1.16 is less than the F-value of 6.03, one may assume comparable variances.

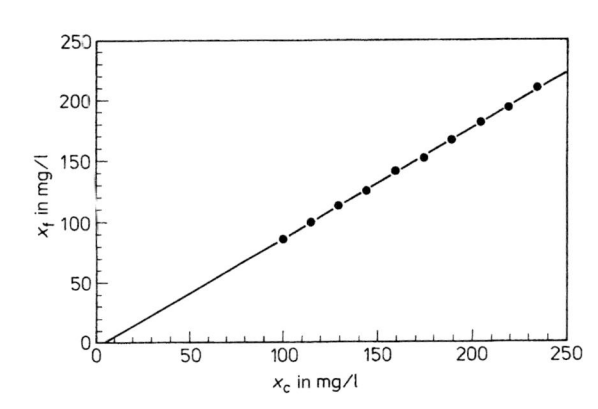

Fig. A10 Recovery function for the digestion step.

Testing for constant systematic errors:

The confidence interval for the axis intercept is

$CI(a_f) = a_f \pm t\,(f = 8, P = 95\%, \text{two-sided}) \cdot s_{a_f}$ $\quad$ (where $t\,(f, P) = 2.306$)
$ = -3.7455$ mg/l $\pm\, 2.306 \cdot 1.499$ mg/l

from which one obtains -7.202 mg/l $\leq a_f \leq -0.289$ mg/l.

This confidence interval does not include the value 0 mg/l; one may therefore assume a constant systematic error.

Testing for proportional systematic errors:

The confidence interval for the slope is

$$CI(bf) = b_f \pm t\,(f = 8, P = 95\%, \text{two-sided}) \cdot s_{b_f} \quad (\text{where } t\,(f, P) = 2.306)$$
$$= 0.898 \pm 2.306 \cdot 0.00867$$

from which one obtains $0.878 \leq b_f \leq 0.918$.

Since the "expected" value $b_f = 1$ is not included, a proportional systematic error exists. The sample digestion step must be improved.

Testing the analytical process for matrix influence:

An actual sample is spiked ten times by the addition of a standard substance. The amounts of standard added were calculated such that the concentration increases were between 100 and 235 mg/l.

No.	$x_{\text{calib.}} = x$	$x_{\text{found}} = y$	x^2	y^2	$x \cdot y$
1	100	183	1.00000E4	3.34890E4	1.83000E4
2	115	197	1.32250E4	3.88090E4	2.26550E4
3	130	212	1.69000E4	4.49440E4	2.75600E4
4	145	225	2.10250E4	5.06250E4	3.26250E4
5	160	243	2.56000E4	5.90490E4	3.88800E4
6	175	256	3.06250E4	6.55360E4	4.48000E4
7	190	272	3.61000E4	7.39840E4	5.16800E4
8	205	288	4.20250E4	8.29440E4	5.90400E4
9	220	302	4.84000E4	9.12040E4	6.64400E4
10	235	320	5.52250E4	1.02400E5	7.52000E4
Σ	1675	2498	2.99125E5	6.42984E5	4.37180E5

Linear regression:

Q_{xx}:   18562.5000 $(\text{mg/l})^2$
Q_{yy}:   18983.6000 $(\text{mg/l})^2$
Q_{xy}:   18765.0000 $(\text{mg/l})^2$
mean: $\bar{x}_{\text{added}} = 167.5$ mg/l
mean: $\bar{x}_{\text{found}} = 249.8$ mg/l
slope: $b_f = 1.0109$
axis intercept: $a_f = 80.4727$ mg/l
residual standard deviation: $s_{y_f} = 1.318$ mg/l
standard deviation of the slope: $s_{b_f} = 0.00967$
standard deviation of the axis intercept: $s_{a_f} = 1.672$ mg/l

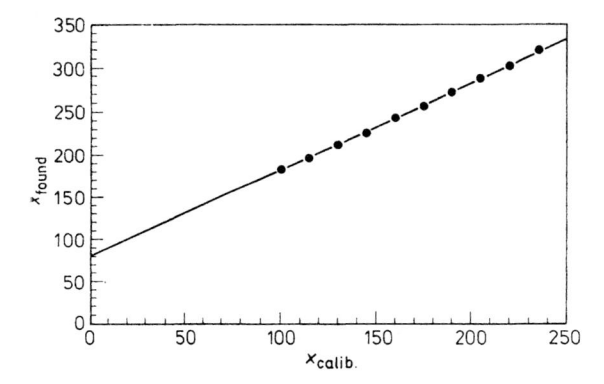

[Figure: plot of x_{found} vs $x_{calib.}$, linear trend from ~80 to ~320]

Fig. A11 Use of the recovery function to test for a matrix influence.

Neither an increase in the imprecision nor a proportional systematic error can be ascertained; testing the axis intercept may not be applied in this case since the substance XYZ is already present in the actual sample (perhaps at approximately 80 mg/l?).

A1.2.9
Testing Analytical Results for Temporal Stability

Using the first calibration function, an actual sample was analyzed at the beginning and end of each workday (day no. i) for each of $N = 20$ days. Both results (x_{i1} and x_{i2}) were entered in the following table; the series mean, $\bar{x}_i$, and the squared sum of the deviation, Q_i, were calculated for each individual series:

$$\bar{x}_i = \frac{1}{2} \cdot (x_{i1} + x_{i2})$$

$$Q_i = (x_{i1} - \bar{x}_i)^2 + (x_{i2} - \bar{x}_i)^2$$

For the subsequent variance analysis, the following were determined:

a) Degrees of freedom

$$f_b = N - 1 = 19$$
$$f_w = 2 \cdot N - N = N = 20$$
$$f_t = f_b + f_w = 39$$

| Series no. | 1st result | 2nd result | Mean | Sum of the squares |
i	x_{i1}	x_{i2}	$\bar{x}_i$	Q_i
1	133.3	133.2	133.25	0.005
2	137.1	136.7	136.90	0.080
3	130.2	130.3	130.25	0.005
4	135.3	135.2	135.25	0.005
5	136.6	136.4	136.50	0.020
6	136.8	136.6	136.70	0.020
7	134.3	133.9	134.10	0.080
8	137.4	136.8	137.10	0.180
9	132.6	132.2	132.40	0.080
10	135.4	135.2	135.30	0.020
11	138.4	138.5	135.45	0.005
12	138.2	137.4	137.80	0.320
13	133.3	132.9	133.10	0.080
14	135.3	134.1	134.70	0.720
15	132.9	132.9	132.90	0.000
16	137.1	137.9	137.50	0.320
17	136.0	136.3	136.15	0.045
18	135.6	136.5	136.05	0.405
19	138.3	137.9	138.10	0.080
20	137.0	136.0	136.50	0.500
			$\bar{\bar{x}} = 135.45$	$Q_w = 2.970$

b) Sums of the squares

$$Q_b = \sum_{i=1}^{20} (\bar{x}_i - \bar{\bar{x}})^2 = 181.690 \text{ mg/l}$$

$$Q_w = \sum_{i=1}^{20} Q_i = 2.970 \text{ mg/l}$$

$$Q_t = Q_b + Q_w = 184.660 \text{ mg/l}$$

c) Variances

$$s_b^2 = \frac{Q_b}{f_b} = 9.5626 \text{ (mg/l)}^2$$

$$s_w^2 = \frac{Q_w}{f_w} = 0.1485 \text{ (mg/l)}^2$$

$$s_t^2 = \frac{Q_t}{f_t} = 4.7349 \text{ (mg/l)}^2$$

d) Grand average

$$\bar{\bar{x}} = \frac{1}{20} \cdot \sum_{i=1}^{20} \bar{x}_i = 135.45 \text{ mg/l}$$

e) Total standard deviation

$$s_t = \sqrt{\frac{1}{39} \sum_{i=1}^{20} \sum_{j=1}^{2} (x_{ij} - \bar{\bar{x}})^2} = 2.176 \text{ mg/l}$$

	Degree of freedom	Sum of the squares	Variance	Standard deviation
Between:	$f_b = 19$	$Q_b = 181.690$	$s_b^2 = 9.5626$	$s_b = \sqrt{s_b^2} = 3.092$ mg/l
Within:	$f_w = 20$	$Q_w = 2.970$	$s_w^2 = 0.1485$	$s_w = \sqrt{s_w^2} = 0.385$ mg/l
Total:	$f_t = 39$	$Q_t = 184.660$	$s_t^2 = 4.7349$	$s_t = \sqrt{s_t^2} = 2.176$ mg/l

The variance F-test yielded a test value of

$$TV = \frac{s_b^2}{s_w^2} = \frac{9.5626}{0.1485} = 63.395$$

Since TV is larger than $F(f_b, f_w, P = 99\%) = 2.94$, the total imprecision is unusually heavily influenced by the time for which the analysis was performed.

This is also evident from the graph of the individual values (Figure A12): the two results of the individual series lie relatively near to one another compared with the total distribution of all data.

There is clearly no drifting of the analytical results over a time period of 20 days. The analytical process must be submitted to testing; if necessary, the calibration function of each series must be determined at the beginning of each series.

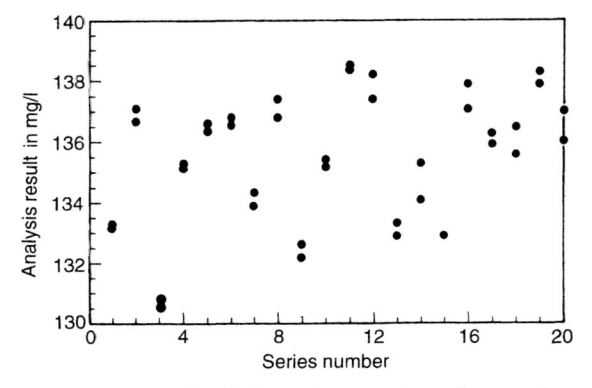

Fig. A12 Results of 20 duplicate determinations of an actual sample in a time frame of 20 days.

A1.2.10 Trend Test

A control sample was analyzed once daily for 21 days.

Series no.	Result x_i	$(x_i - \bar{x})^2$	$(x_i - x_{i-1})^2$
1	1.043	0.01753733	
2	1.071	0.01090533	0.00004398
3	1.081	0.00891676	0.00000395
4	1.078	0.00949233	0.00000033
5	1.108	0.00454661	0.00002446
6	1.162	0.00018033	0.00001906
7	1.146	0.00086604	0.00000047
8	1.171	0.00001961	0.00000072
9	1.147	0.00080818	0.00000062
10	1.161	0.00020818	0.00000036
11	1.180	0.00002090	0.00000004
12	1.209	0.00112704	0.00000122
13	1.205	0.00087447	0.00000006
14	1.184	0.00007347	0.00000064
15	1.223	0.00226304	0.00000479
16	1.211	0.00126533	0.00000100
17	1.270	0.00894376	0.00005896
18	1.219	0.00189847	0.00004964
19	1.274	0.00971633	0.00006112
20	1.298	0.01502376	0.00002817
21	1.243	0.00456590	0.00010937
Sum	24.684	$s^2 = 0.09925314$	$\Delta^2 = 0.00040896$

Mean $\bar{x} = 1.175$ mg/l

The mean

$$\bar{x} = \frac{1}{21} \sum_{i=1}^{21} x_i = 1.175 \text{ mg/l}$$

and the square of the deviations $(x_i - \bar{x})^2$ and $(x_i - x_{i-1})^2$ (from the second analysis on) were calculated. The square of the deviations was summed by rows and the test value

$$TV = \frac{\Delta^2}{s^2}$$

was calculated.

$$TV = \frac{0.00040896}{0.09925314} = 0.00412042$$

The trend, which is visually recognizable in Figure A13, was statistically verified by $TV < 1.0601$. The cause of the trend could be an error in the preservation of the control sample.

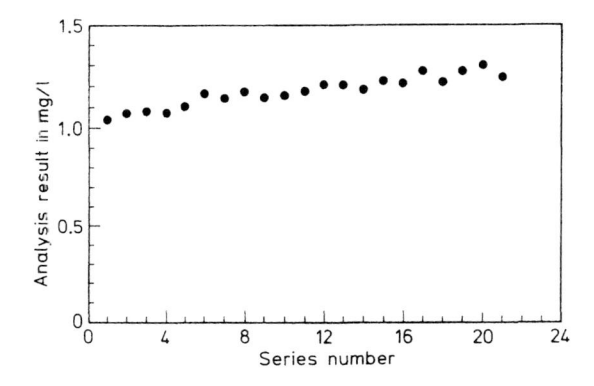

Fig. A13 Analytical results of the control sample over 21 days.

A1.2.11 Practice Phase: Checking the Analysis Quality Achieved Based on the Process Standard Deviation

First calibration:

No.	x in mg/l	y in abs.	x^2	y^2	$x \cdot y$
1	100	0.751	1.00000E4	5.64001E-1	7.51000E1
2	115	0.807	1.32250E4	6.51249E-1	9.28050E1
3	130	0.946	1.69000E4	8.94916E-1	1.22980E2
4	145	1.002	2.10250E4	1.00400E0	1.45290E2
5	160	1.141	2.56000E4	1.30188E0	1.82560E2
6	175	1.197	3.06250E4	1.43281E0	2.09475E2
7	190	1.336	3.61000E4	1.78490E0	2.53840E2
8	205	1.391	4.20250E4	1.93488E0	2.85155E2
9	220	1.531	4.84000E4	2.34396E0	3.36820E2
10	235	1.587	5.52250E4	2.51857E0	3.72945E2
Σ	1675	11.689	2.99125E5	1.44312E1	2.07697E3

Linear regression:

Q_{xx}: 18562.5000 $(mg/l)^2$
Q_{yy}: 0.7679 $abs.^2$
Q_{xy}: 119.0625 abs. · mg/l
mean $\bar{x}$: 167.5000 mg/l
mean $\bar{y}$: 1.1689 abs.
slope b: 0.0064 abs./(mg/l)

Appendix 1

axis intercept a: 0.0945 abs.
residual standard deviation s_y: 0.023 abs.
process standard deviation s_{xo}: 3.577 mg/l
process variation coefficient V_{xo}: 2.14 %
t ($f = 8$, $P = 95\%$, one-sided): 1.860
auxiliary value y_a: 0.1635 abs.
test value x_a: 21.50 mg/l

The entire calibration was carried out three more times, whereby the practice effect is reflected in the decrease of the process standard deviation.

Testing the process standard deviation:

Expected value		s_{xo} expected =	1.096 in mg/l	$(s_{xo_i}/s_{xo\ exp})^2$
Calibration number 1	s_{xo_1}	=	3.577	10.652
2	s_{xo_2}	=	2.876	6.884
3	s_{xo_3}	=	2.392	4.764
4	s_{xo_4}	=	1.430	1.703

The threshold value of the F-distribution F ($f_1 = 8$, $f_2 = 8$, $P = 95\%$) is 3.44.

Consequently, one may regard the expected process standard deviation as having been achieved with the fourth calibration.

A1.3 Phases II and III: Control Charts

For routine quality assurance, several control charts for different control samples must be maintained simultaneously:

Type of control sample			Control chart				
	$\bar{x}$	Blank	RR	Range	s	Difference	Cusum
Blank sample		×					
Standard solution 180 mg/l	×		×				×
Actual sample: duplicate determination				×	×	×	
Actual sample + spiking with 60 mg/l			×				

A preliminary period of 25 series is maintained for all control charts. The additional effort required for quality assurance in each series consists of four analyses:

- one blank value
- one standard solution
- one second analysis of the actual sample at the end of the series
- one spiked actual sample

Note: Multiple analyses of the blank value sample are not considered necessary for an *s*-control chart; a final decision should be made based on the requirements of the routine analyses to follow.

A1.3.1 Blank Value Control Chart

Series no.	Blank value in abs.
1	0.089
2	0.080
3	0.091
4	0.083
5	0.079
6	0.054
7	0.081
8	0.082
9	0.065
10	0.073
11	0.080
12	0.073
13	0.081
14	0.089
15	0.079
16	0.076
17	0.069
18	0.083
19	0.076
20	0.075
21	0.083
22	0.097
23	0.068
24	0.066
25	0.079
Sum	1.951
Mean	0.078
Standard deviation	0.009

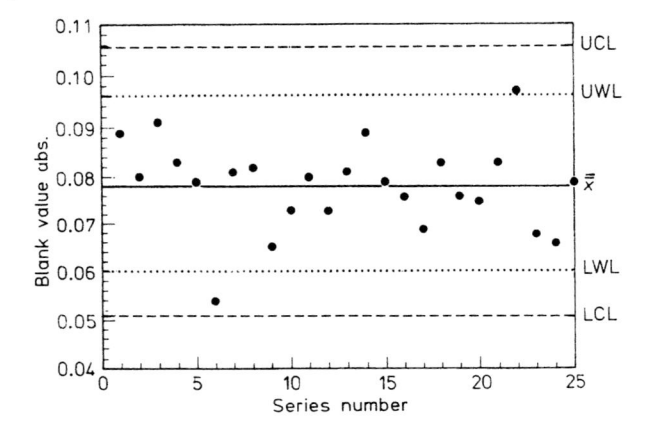

Fig. A14 Blank value control chart.

A1.3.2 $\bar{x}$-Chart for Standard Solutions

Series no.	Results in mg/l	Deviation in mg/l	Normalized deviation
1	179.3	−1.00	−0.56
2	179.9	−0.40	−0.22
3	182.6	2.30	1.28
4	180.4	0.10	0.06
5	180.8	0.50	0.28
6	180.9	0.60	0.33
7	179.8	−0.50	−0.28
8	183.7	3.40	1.89
9	178.2	−2.10	−1.17
10	180.9	0.60	0.33
11	178.9	−1.40	−0.78
12	184.5	4.20	2.33
13	181.6	1.30	0.72
14	180.8	0.50	0.28
15	179.0	−1.30	−0.72
16	179.5	−0.80	−0.44
17	178.3	−2.00	−1.11
18	177.7	−2.60	−1.44
19	179.3	−1.00	−0.56
20	179.1	−1.20	−0.67
21	182.1	1.80	1.00
22	182.9	2.60	1.44
23	180.3	0.00	0.00
24	178.4	−1.90	−1.06
25	179.0	−1.30	−0.72

Sum	4507.9
Mean	180.3
Standard deviation	1.8

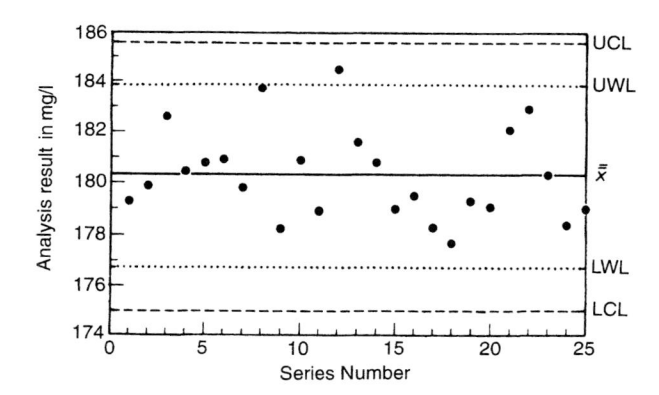

Fig. A15 $\bar{x}$-chart.

Alternatively, the analytical results may also be entered into a normalized diagram. For this, the deviations of the individual values are calculated and divided by the standard deviation:

$$\text{normalized deviation} = \frac{\text{individual value} - \text{mean}}{\text{standard deviation}}$$

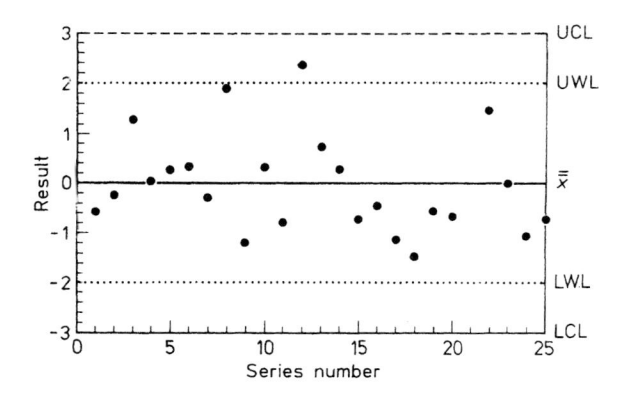

Fig. A16 Normalized $\bar{x}$-chart.

A1.3.3 Recovery Rate (*RR*) Control Chart

The results for the standard sample were also used for an *RR* control chart (Figure A17):

$$RR = \frac{\text{result}}{180 \text{ mg/l}} \cdot 100\%$$

The central line was defined as a 100% recovery rate.

Appendix 1

Series no	Result in mg/l	RR in %
1	179.3	99.6
2	179.9	99.9
3	182.6	101.4
4	180.4	100.2
5	180.8	100.4
6	180.9	100.5
7	179.8	99.9
8	183.7	102.1
9	178.2	99.0
10	180.9	100.5
11	178.9	99.4
12	184.5	102.5
13	181.6	100.9
14	180.8	100.4
15	179.0	99.4
16	179.5	99.7
17	178.3	99.1
18	177.7	98.7
19	179.3	99.6
20	179.1	99.5
21	182.1	101.2
22	182.9	101.6
23	180.3	100.2
24	178.4	99.1
25	179.0	99.4
Sum	4507.9	2504.4
Mean	180.3	100.2
Standard deviation	1.8	1.0

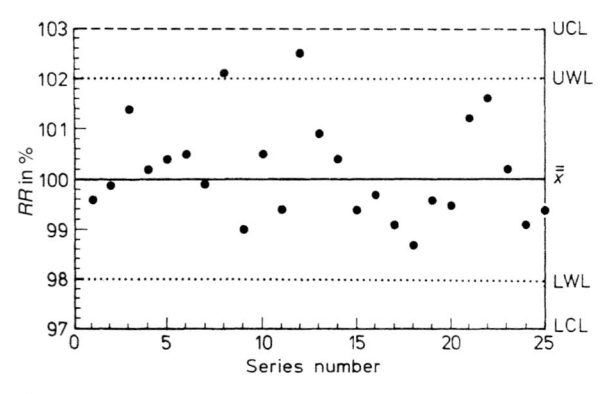

Fig. A17 *RR* control chart for a standard of concentration of 180 mg/l.

A1.3.4 Verifying Precision by Means of *R*-Charts and *s*-Charts

One of the first actual samples of each series is divided into three portions and is analyzed as usual at both the beginning and the end of the series (the third aliquot is needed for the spiking analysis).

Results [mg/l]:

Series no.	First	Second	Range
1	143.4	141.0	2.3
2	140.6	137.1	3.4
3	175.8	175.2	0.5
4	165.0	164.9	0.2
5	169.8	167.8	2.0
6	171.0	172.9	1.9
7	185.8	183.8	2.1
8	173.4	173.7	0.3
9	167.8	165.5	2.3
10	171.7	172.5	0.8
11	168.2	173.9	5.7
12	155.2	157.4	2.2
13	177.1	176.7	0.3
14	197.8	197.8	0.0
15	134.4	135.6	1.2
16	188.5	187.4	1.1
17	158.7	158.3	0.4
18	175.5	176.1	0.6
19	154.1	154.6	0.5
20	164.0	162.9	1.1
21	147.1	148.5	1.4
22	160.1	161.6	1.5
23	186.8	184.6	2.2
24	124.1	123.7	0.4
25	184.8	182.1	2.6
Sum	4140.7	4135.6	37.0
Mean	165.6	165.4	1.49
Standard deviation	18.0	17.9	1.3

The individual ranges are calculated as the difference between the two analytical results:

Range = |first result − second result|

The control limits obtained are:

$\text{UCL} = D_4 \cdot \bar{R} = 3.267 \cdot \bar{R} = 4.87$
$\text{LCL} = D_3 \cdot \bar{R} = 0 \cdot \bar{R} = 0$

The noticeably wider range on the 11th day (which exceeded the control limit) could be identified as a "Monday effect": the photometer was zeroed early in the morning at a very low room temperature. As a consequence, the timing of the heating in the laboratory building was modified.

The corresponding *s-control chart* provided analogous results. The standard deviation of the duplicate determination was calculated according to:

$$s_i = \sqrt{\frac{1}{n_i - 1} \sum_{j=1}^{n_i} (x_{ij} - \bar{x}_i)^2}$$

Series no.	Result 1 in mg/l	Result 2 in mg/l	Range in mg/l	s_i in mg/l	s_i^2 in mg/l²
1	143.4	141.0	2.3	1.70	2.8900
2	140.6	137.1	3.4	2.47	6.1009
3	175.8	175.2	0.5	0.42	0.1764
4	165.0	164.9	0.2	0.07	0.0049
5	169.8	167.8	2.0	1.41	1.9881
6	171.0	172.9	1.9	1.34	1.7956
7	185.8	183.8	2.1	1.41	1.9881
8	173.4	173.7	0.3	0.21	0.0441
9	167.8	165.5	2.3	1.63	2.6569
10	171.7	172.5	0.8	0.57	0.3249
11	168.2	173.9	5.7	4.03	16.2409
12	155.2	157.4	2.2	1.56	2.4336
13	177.1	176.7	0.3	0.28	0.0784
14	197.8	197.8	0.0	0.00	0.0000
15	134.4	135.6	1.2	0.85	0.7225
16	188.5	187.4	1.1	0.78	0.6084
17	158.7	158.3	0.4	0.28	0.0784
18	175.5	176.1	0.6	0.42	0.1764
19	154.1	154.6	0.5	0.35	0.1225
20	164.0	162.9	1.1	0.78	0.6084
21	147.1	148.5	1.4	0.99	0.9801
22	160.1	161.6	1.5	1.06	1.1236
23	186.8	184.6	2.2	1.56	2.4336
24	124.1	123.7	0.4	0.28	0.0784
25	184.8	182.1	2.6	1.91	3.6481
Sum	4140.7	4135.6	37.0		47.3032
Mean	165.6	165.4	1.49		$s_w = 1.3755$
Std. dev.	18.0	17.9	1.3		
Degree of freedom					$f = \Sigma f_i = 25$

The central line was

$$s_w = \sqrt{\frac{\Sigma s_i^2}{25}} = 1.375 \text{ mg/l}$$

where the control limits (for $\alpha = 1\%$; χ^2 from Table 2.8) were

$$\text{UCL} = s_w \cdot \sqrt{\frac{1}{1} \cdot \chi^2\left(1, \frac{\alpha}{2}\right)} = 1.375 \cdot \sqrt{1 \cdot 7.879} = 3.852 \text{ mg/l}$$

$$\text{LCL} = s_w \cdot \sqrt{\frac{1}{1} \cdot \chi^2\left(1, 1 - \frac{\alpha}{2}\right)} = 1.375 \cdot \sqrt{1 \cdot 0.0000393} = 0.00862 \text{ mg/l}$$

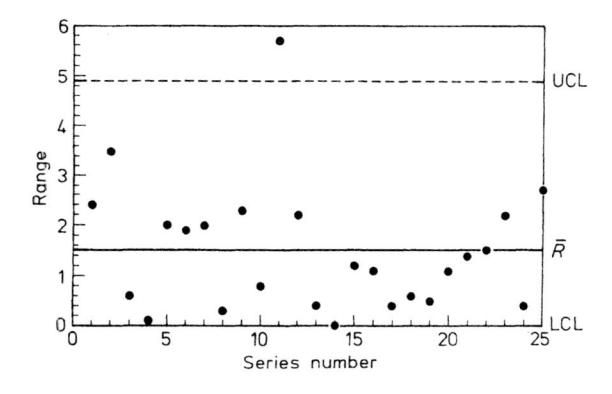

Fig. A18 R-chart.

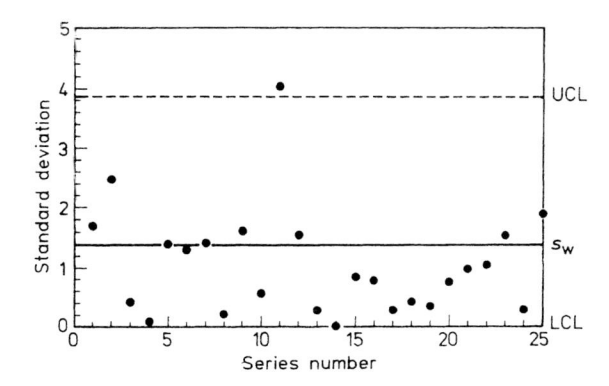

Fig. A19 s-control chart for duplicate determinations of actual samples.

A1.3.5 Testing for Series-Internal Drift

Instead of the ranges (= absolute value of the difference between the two results of the duplicate determination), here the difference between the second and first results is determined while incorporating the sign:

difference = second result − first result

Except for the previously known irregularity in the 11th series, no significant deviations could be detected from the graphical representation of the differences in the control chart (Figure A20). In particular, the mean difference is near zero; the individual differences are randomly distributed with changing signs, and so a drift of the results in the direction of systematically smaller or larger values between the beginning and end of the series can be ruled out.

Appendix 1

Series no.	Results in mg/l		Difference in mg/l
	First	Second	
1	143.4	141.0	−2.4
2	140.6	137.1	−3.5
3	175.8	175.2	−0.6
4	165.0	164.9	−0.1
5	169.8	167.8	−2.0
6	171.0	172.9	1.9
7	185.8	183.8	−2.0
8	173.4	173.7	0.3
9	167.8	165.5	−2.3
10	171.7	172.5	0.8
11	168.2	173.9	5.7
12	155.2	157.4	2.2
13	177.1	176.7	−0.4
14	197.8	197.8	0.0
15	134.4	135.6	1.2
16	188.5	187.4	−1.1
17	158.7	158.3	−0.4
18	175.5	176.1	0.6
19	154.1	154.6	0.5
20	164.0	162.9	−1.1
21	147.1	148.5	1.4
22	160.1	161.6	1.5
23	186.8	184.6	−2.2
24	124.1	123.7	−0.4
25	184.8	182.1	−2.7
Sum			−5.1
Mean			−0.2
Standard deviation			1.96

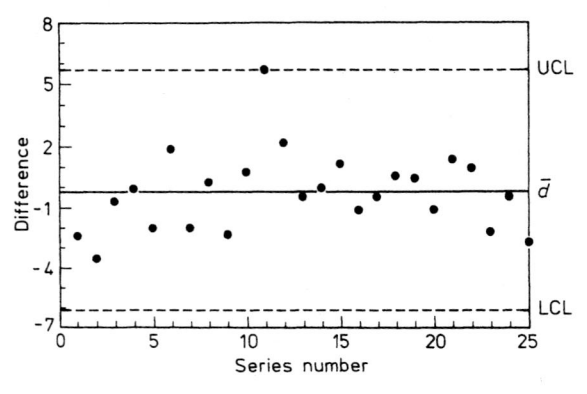

Fig. A20 Difference control chart.

A1.3.6 RR-Control Chart by Addition of a Standard

The third portion from the actual sample was spiked in such a way that the content of XYZ increased by 60 mg/l:

sample volume: $V_S = 1\,l$

standard addition: $V_A = 0.5$ ml; concentration $c_A = 120.061$ g/l

spiking concentration: $c_S = \dfrac{V_A \cdot c_A}{V_S + V_A} = \dfrac{0.0005\,l \cdot 120.061\,g/l}{1\,l + 0.0005\,l} = 0.060$ g/l

Series no.	Analytical results		RR in %
	before	after	
1	143.4	202.2	98.0
2	140.6	212.7	120.3
3	175.8	228.6	88.1
4	165.0	218.1	88.5
5	169.8	228.8	98.4
6	171.0	244.2	122.0
7	185.8	246.5	101.1
8	173.4	226.2	88.0
9	167.8	231.3	105.9
10	171.7	234.4	104.5
11	168.2	229.9	102.7
12	155.2	219.8	107.6
13	177.1	245.7	114.4
14	197.8	264.4	110.9
15	134.4	189.8	92.3
16	188.5	244.6	93.6
17	158.7	222.6	106.6
18	175.5	232.7	95.4
19	154.1	210.7	94.3
20	164.0	222.0	96.7
21	147.1	211.9	108.0
22	160.1	221.3	102.1
23	186.8	241.9	91.9
24	124.1	179.8	92.9
25	184.8	244.2	99.0
Sum	4140.7	5654.3	2523.2
Mean	165.6	226.2	100.9
Standard deviation	18.0	18.8	9.4

UCL $= \overline{RR} + 3 \cdot s = 100.9\% + 3 \cdot 9.4\% = 129.1\%$
UWL $= \overline{RR} + 2 \cdot s = 100.9\% + 2 \cdot 9.4\% = 119.7\%$
LWL $= \overline{RR} - 2 \cdot s = 100.9\% - 2 \cdot 9.4\% = 82.1\%$
LCL $= \overline{RR} - 3 \cdot s = 100.9\% - 3 \cdot 9.4\% = 72.7\%$

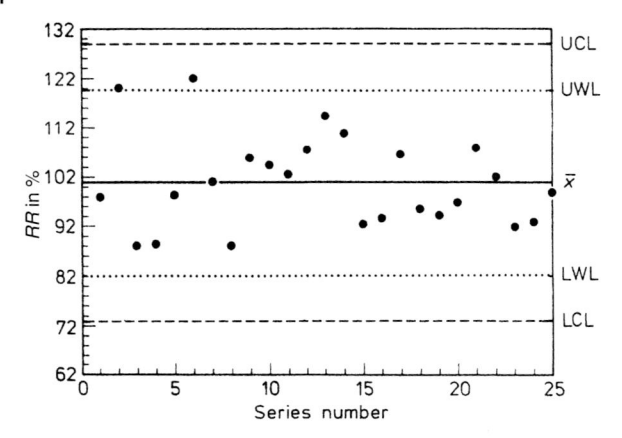

Fig. A21 RR control chart for an actual spiked sample.

A1.3.7 Cusum Chart

To determine systematic errors, the results for the standard sample were recalculated for a cusum control chart. The expected concentration:

$k = 180$ mg/l

was given as a reference value, for which the current deviation was calculated:

$dev = result - expected\ value$

The cumulative deviation sums were determined by the summation of all individual deviations:

$$cusum_i = \sum_{j=1}^{i} dev_j$$

Example: series no. 4

$$cusum = dev_1 + dev_2 + dev_3 + dev_4$$
$$= -0.70 + (-0.10) + 2.60 + 0.40$$
$$= 2.20\ \text{mg/l}$$

The standard deviation for the preliminary period was $s = 1.8$ mg/l.
The cusum value was normalized by dividing by the standard deviation

$$normalized = \frac{cusum}{s} = \frac{cusum}{1.8}$$

The cusum chart was then drawn using a scale of $w = 1 \cdot s$ or $w = 2 \cdot s$.
The normalized cusum values were entered and a test V-mask was constructed (Figure A22).

Series no.	Result in mg/l	Dev. in mg/l	Cusum in mg/l	Normalized
1	179.3	−0.70	−0.70	−0.39
2	179.9	−0.10	−0.80	−0.45
3	182.6	2.60	1.80	1.01
4	180.4	0.40	2.20	1.24
5	180.8	0.80	3.00	1.69
6	180.9	0.90	3.90	2.19
7	179.8	−0.20	3.70	2.08
8	183.7	3.70	7.40	4.16
9	178.2	−1.80	5.60	3.15
10	180.9	0.90	6.50	3.65
11	178.9	−1.10	5.40	3.03
12	184.5	4.50	9.90	5.56
13	181.6	1.60	11.50	6.46
14	180.8	0.80	12.30	6.91
15	179.0	−1.00	11.30	6.35
16	179.5	−0.50	10.80	6.07
17	178.3	−1.70	9.10	5.11
18	177.7	−2.30	6.80	3.82
19	179.3	−0.70	6.10	3.43
20	179.1	−0.90	5.20	2.92
21	182.1	2.10	7.30	4.10
22	182.9	2.90	10.20	5.73
23	180.3	0.30	10.50	5.90
24	178.4	−1.60	8.90	5.00
25	179.0	−1.00	7.90	4.44

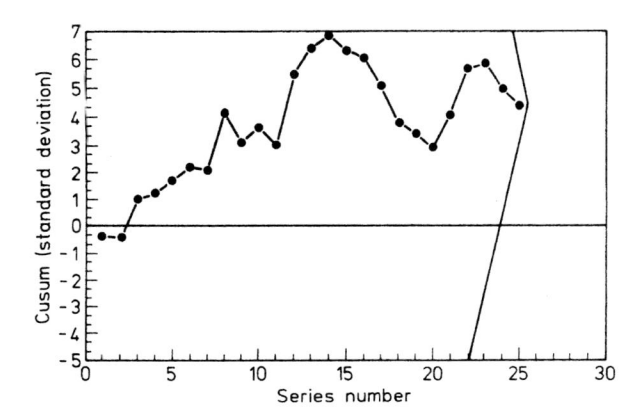

Fig. A22 Cusum chart for the standard 180 mg/l.

Appendix 1

In the routine analysis to follow, the cusum chart was maintained for the 180 mg/l control sample:

Routine analysis: cusum

Series no.	Result in mg/l	Deviation in mg/l	Normalized deviation	Cusum of the normalized deviation
1	179.9	−0.1	−0.06	−0.06
2	180.9	0.9	0.50	0.44
3	176.4	−3.6	−2.00	−1.56
4	179.7	−0.3	−0.17	−1.72
5	182.3	2.3	1.28	−0.44
6	178.9	−1.1	−0.61	−1.06
7	177.5	−2.5	−1.39	−2.44
8	178.2	−1.8	−1.00	−3.44
9	180.5	0.5	0.28	−3.17
10	181.5	1.5	0.83	−2.33
11	181.4	1.4	0.78	−1.56
12	179.2	−0.8	−0.44	−2.00
13	181.7	1.7	0.94	−1.06
14	181.0	1.0	0.56	−0.50
15	176.0	−4.0	−2.22	−2.72
16	178.1	−1.9	−1.06	−3.78
17	179.3	−0.7	−0.39	−4.17
18	179.3	−0.7	−0.39	−4.56
19	179.2	−0.8	−0.44	−5.00
20	177.1	−2.9	−1.61	−6.61
21	177.6	−2.4	−1.33	−7.94
22	175.4	−4.6	−2.56	−10.50 [a]
23	181.6	1.6	0.89	0.89 [b]
24	180.6	0.6	0.33	1.22
25	183.5	3.5	1.94	3.16

[a] corrective measures necessary,
[b] begin again.

The cusum chart was tried out using two different axis scales (Figures A23 and A24):

scale factor	$w = 1 \cdot s$	$w = 2 \cdot s$
smallest deviation that should be detected	$D = 1.5 \cdot s$	$\left(\delta = \dfrac{D}{s} = 1{,}5\right)$
probability of error	$\alpha = 0.0027$	
distance, d, from the V-mask	$d = \dfrac{-2}{\delta^2} \cdot \ln \alpha$	
	$= \dfrac{-2}{1.5^2} \cdot \ln 0.0027$	
	$d = 5.2573$	$d = 5.2573$

angle of the V-mask $\quad \theta = \arctan\left(\dfrac{D}{2 \cdot w}\right)$

$\theta = \arctan\left(\dfrac{1.5s}{2 \cdot 1s}\right) \qquad \theta = \arctan\left(\dfrac{1.5s}{2 \cdot 2s}\right)$

$\theta = 36.87° \qquad\qquad\qquad \theta = 20.56°$

(The mask parameters correspond approximately to ARL curve no. XI in Figure 2-22).

The systematic error in the mean was obviously present from series no. 15 onwards and it was detected in the 22nd series.

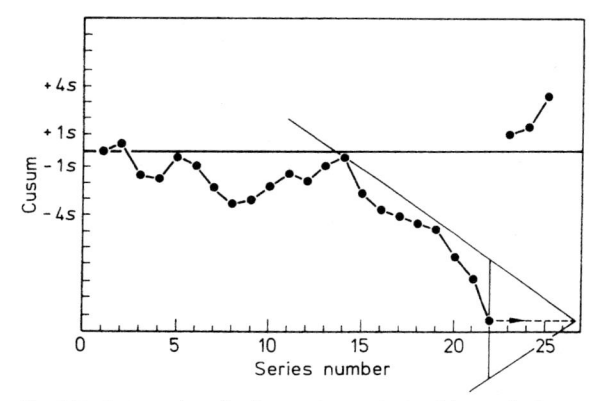

Fig. A23 Cusum chart for the routine analysis with a scale factor of $w = 1s$.

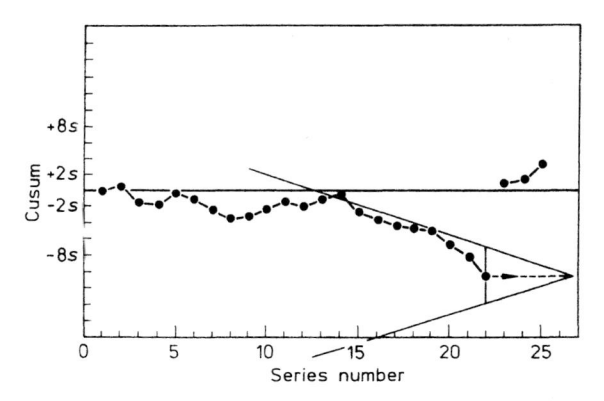

Fig. A24 Cusum chart for the routine analysis with a scale factor of $w = 2s$.

A1.3.8 Equivalency

a) Difference methods

Parameter: Magnesium
Reference procedure: ICP-OES
Comparison procedure: Ion chromatography

Analytical results from the reference and comparison procedures:

i	x_{Ri} in mg/l	x_{CPi} in mg/l	$D_i = x_{Ri} - x_{CPi}$ in mg/l	i	x_{Ri} in mg/l	x_{CPi} in mg/l	$D_i = x_{Ri} - x_{CPi}$ in mg/l
1	10.50	10.77	−0.27	16	24.80	21.00	3.80
2	10.50	10.99	−0.49	17	18.60	19.54	−0.94
3	9.05	9.39	−0.34	18	27.00	22.61	4.39 [a)]
4	9.23	9.43	−0.20	19	24.91	25.58	−0.67
5	19.20	17.50	1.70	20	24.93	25.66	−0.73
6	15.30	16.94	−1.64	21	24.95	25.69	−0.74
7	13.02	12.90	0.12	22	24.96	25.71	−0.75
8	14.51	14.69	−0.18	23	25.01	24.29	0.72
9	14.55	14.28	0.27	24	28.80	25.79	3.01
10	13.08	13.02	0.06	25	34.25	34.41	−0.17
11	23.00	18.86	4.14	26	34.26	34.50	−0.24
12	18.80	17.00	1.80	27	30.90	30.96	−0.06
13	14.58	15.42	−0.84	28	26.60	27.73	−1.13
14	17.20	14.00	3.20	29	28.00	25.09	2.91
15	14.85	14.20	0.65	30	31.08	28.13	2.95

a) Suspected outlier value D^*.

$$\bar{D} = \frac{\sum D_i}{N} = 0.678 \text{ mg/l}$$

$$s_D = \sqrt{\frac{\sum (D_i - \bar{D})^2}{N-1}} = 1.751 \text{ mg/l}$$

Outlier test:

$$TV = \frac{|D^* - \bar{D}|}{s_D} = \frac{|4.39 - 0.68|}{1.751} = 2.120$$

$N = 30$

$rM\ (f = N;\ P = 95\%) = 2.745$

$TV < 2.745$; therefore, there are no outliers

Joint t-test:

$$TV = \frac{|\bar{D}|}{s_D} \cdot \sqrt{N} = 2.12$$

$t\ (f = N - 1;\ P = 99\%) = 2.756$

$TV < 2.756$; therefore, there is *no* significant difference; the results are equivalent.

Magnesium

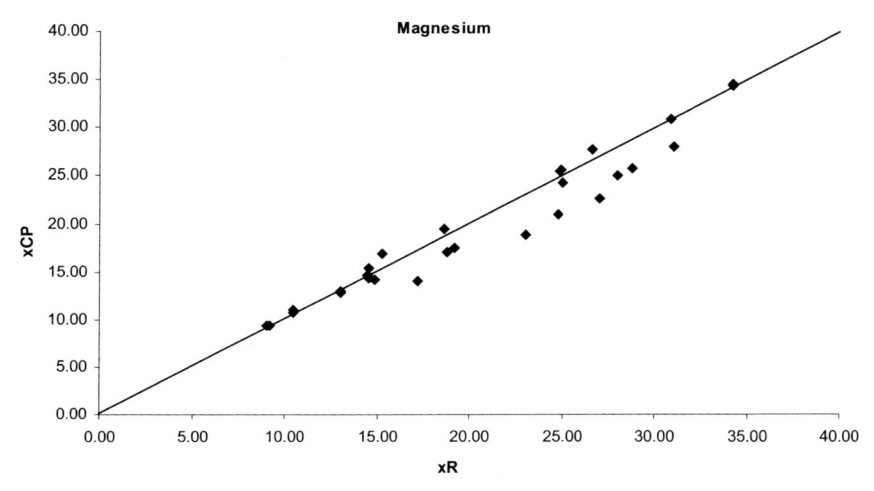

Fig. A25 Equivalency using the difference method.

b) Orthogonal regression

Parameter: AOX
Reference procedure: DEV H 14
Comparison procedure: DEV H 22 (highly saline water)

Analytical results from the reference and comparison procedures:

i	x_{Ri} in mg/l	x_{CPi} in mg/l	$Q_i = \dfrac{x_{Ri}}{x_{CPi}}$	i	x_{Ri} in mg/l	x_{CPi} in mg/l	$Q_i = \dfrac{x_{Ri}}{x_{CPi}}$
1	0.27	0.50	1.85	19	1.80	2.00	1.11
2	0.29	0.60	2.07	20	2.00	2.10	1.05
3	0.45	1.00	2.22	21	2.10	2.60	1.24
4	0.48	0.80	1.67	22	2.20	2.90	1.32
5	0.49	1.10	2.24	23	2.30	2.30	1.00
6	0.73	1.10	1.51	24	2.30	2.70	1.17
7	0.74	1.00	1.35	25	2.50	1.60	0.64
8	0.79	1.30	1.65	26	2.60	2.90	1.12
9	0.81	3.80	4.69[a]	27	2.70	3.60	1.33
10	1.00	1.40	1.40	28	2.70	3.30	1.22
11	1.10	1.50	1.36	29	2.80	3.30	1.18
12	1.20	1.80	1.50	30	2.80	3.30	1.18
13	1.30	1.50	1.15	31	3.30	3.60	1.09
14	1.30	2.20	1.69	32	3.40	3.90	1.15
15	1.40	2.00	1.43	33	3.40	4.00	1.18
16	1.60	2.30	1.44	34	3.40	4.00	1.18
17	1.70	1.90	1.12	35	3.50	3.80	1.09
18	1.80	1.60	0.89				

[a] Suspected outlier value pair.

Appendix 1

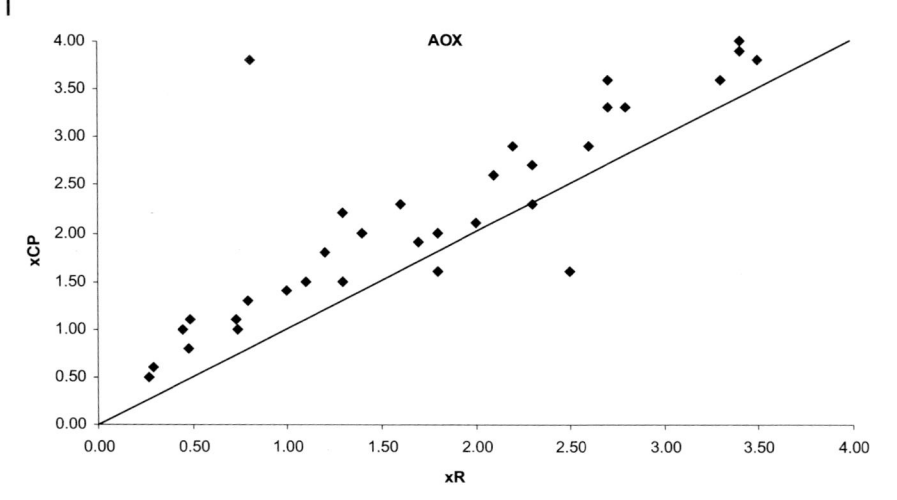

Fig. A26 Equivalency using orthogonal regression.

$$\bar{Q} = \frac{\sum Q_i}{N} = 1.442$$

$$s_Q = \sqrt{\frac{\sum (Q_i - \bar{Q})^2}{N-1}} = 0.6655$$

Outlier test:

$$TV = \frac{|Q^* - \bar{Q}|}{s_Q} = 4.882$$

$rM\,(f = 35;\ P = 95\,\%) = 2.811$

$TV > 2.811$; therefore, there is an outlier.

$N = 34$ (1 outlier removed)

Standard deviation without consideration of the outlier value pair:

$s_R = 1.01$ mg/l $\qquad s_{CP} = 1.06$ mg/l

$$s = \sqrt{\frac{1}{2} \cdot (s_R^2 + s_{CP}^2)} = 1.0356 \text{ mg/l}$$

$$s_{RCP} = \sqrt{\frac{1}{N-1} \sum (x_i - \bar{x}_R)(x_{CPi} - \bar{x}_{CP})} = 1.0193 \text{ mg/l}$$

$$b = \frac{s_{CP}}{s_R} = 1.049$$

$$a = \bar{x}_{CP} - b \cdot \bar{x}_R = 0.294 \text{ mg/l}$$

Test for systematic deviations:

$$X^2 = N \cdot \ln\left(\frac{s^4 - s_{RCP}^4}{s_R^2 s_{CP}^2 - s_{RCP}^4}\right) = 0.8078$$

$X^2 < 3.8$; therefore, there is no proportional systematic deviation.

Joint t-test:

$$\bar{D} = \bar{x}_{CP} - \bar{x}_R = 0.384 \text{ mg/l}$$

$$s_D = \sqrt{\frac{1}{N-1} \sum (D_i - \bar{D})^2} = 0.326 \text{ mg/l}$$

$$TV = \frac{|\bar{x}_{CP} - \bar{x}_R|}{s_D} \sqrt{N} = 6.873$$

$t(f = 33; P = 99\%) = 2.733$

$TV > 2.733$; therefore, there is a constant systematic deviation.

The results are *not* equivalent!

A1.3.9 Standard Addition

The measured value $y_0 = 0.811$ abs. was obtained for an actual sample. The concentration corresponding to this measurement was estimated using the calibration function for aqueous standards:

$$x_0 \approx 100 \text{ mg/l}$$

The spiking concentrations x_S were given as 25, 50, 75, and 100 mg/l:

No.	x_S in mg/l	y in abs.	x_S^2	y^2	$x_S \cdot y$
1	0	0.811	0.00000	6.57721E-1	0.00000
2	25	0.998	6.25000E2	9.96004E-1	2.49500E1
3	50	1.135	2.50000E3	1.28822E0	5.67500E1
4	75	1.331	5.62500E3	1.77156E0	9.98250E1
5	100	1.459	1.00000E4	2.12868E0	1.45900E2
Σ	250	5.734	1.87500E4	6.84219E0	3.27425E2

Linear regression:

Q_{xx}:   6250.0000 (mg/l)^2
Q_{yy}:   0.2664 abs.^{2}
Q_{xy}:   40.7250 (mg/l) · abs.
mean $\bar{x}$: 50.0000 mg/l
mean $\bar{y}$: 1.1468 abs.

Appendix 1

slope b_S:	0.006516 abs./(mg/l)
axis intercept $\hat{y}_0$:	0.8210 abs.
residual standard deviation s_y:	0.0189 abs.
process standard deviation s_{xo}:	2.907 mg/l
process variation coefficient V_{xo}:	5.81%
$t\,(f=3,\,P=95\%,\text{ one-sided})$:	2.353
$t\,(f=3,\,P=95\%,\text{ two-sided})$:	3.182
auxiliary value y_a:	0.877 abs.
test value x_a:	17.31 mg/l
blank value y_B:	0.004 abs.
analytical result $\hat{x}$:	125.4 mg/l
confidence interval: $CI(\hat{x})$	±17.9 mg/l

The result of 125 mg/l lies well above the test value of $x_a = 17.31$ and is therefore valid.

Calculation formulae:

$$x_a = 2 \cdot \frac{y_a - \hat{y}_0}{b_S}$$

where

$$y_a = \hat{y}_0 + s_y \cdot t_{f,P} \cdot \sqrt{\frac{1}{N} + 1 + \frac{\bar{x}^2}{\sum (x_i - \bar{x})^2}}$$

$$\hat{x} = \frac{\hat{y}_0 - y_B}{b_S}$$

$$CI(\hat{x}) = \frac{s_y}{b_S} \cdot t_{f,P} \sqrt{\frac{1}{N} + 1 + \frac{(\hat{y}_0 - \bar{y})^2}{b_S^2 \sum (x_i - \bar{x})^2}}$$

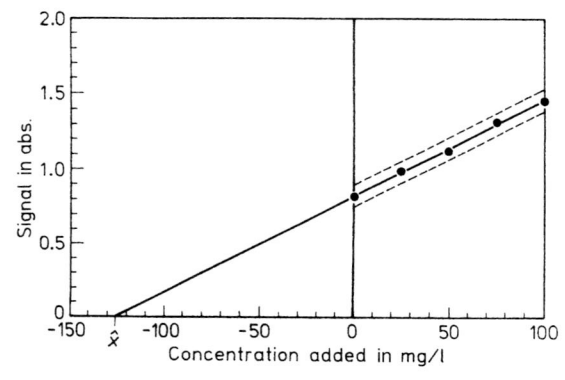

Fig. A27 Standard addition, spiked calibration function with 95% prognosis interval.

Appendix 2

A2
Statistical Tables

A2.1
t-Table

	Double-sided			
f	$P = 90\%$	$P = 95\%$	$P = 99\%$	$P = 99.9\%$
1	6.314	12.706	63.657	636.619
2	2.920	4.303	9.925	31.598
3	2.353	3.182	5.841	12.924
4	2.132	2.776	4.604	8.610
5	2.015	2.571	4.032	6.869
6	1.943	2.447	3.707	5.959
7	1.895	2.365	3.499	5.408
8	1.860	2.306	3.355	5.041
9	1.833	2.262	3.250	4.781
10	1.812	2.228	3.169	4.587
11	1.796	2.201	3.106	4.437
12	1.782	2.179	3.055	4.318
13	1.771	2.160	3.016	4.221
14	1.761	2.145	2.977	4.140
15	1.753	2.131	2.947	4.073
16	1.746	2.120	2.921	4.015
17	1.740	2.110	2.898	3.965
18	1.734	2.101	2.878	3.922
19	1.729	2.093	2.861	3.883
20	1.725	2.086	2.845	3.850
21	1.721	2.080	2.831	3.819
22	1.717	2.074	2.819	3.792
23	1.714	2.069	2.807	3.767
24	1.711	2.064	2.797	3.745
25	1.708	2.060	2.787	3.725
26	1.706	2.056	2.779	3.707
27	1.703	2.052	2.771	3.690
28	1.701	2.048	2.763	3.674
29	1.699	2.045	2.756	3.659
30	1.697	2.042	2.750	3.646
∞	1.645	1.960	2.576	3.291
f	$P = 95\%$	$P = 97.5\%$	$P = 99.5\%$	$P = 99.95\%$
	Single-sided			

Quality Assurance in Analytical Chemistry: Applications in Environmental, Food, and Materials Analysis, Biotechnology, and Medical Engineering, Second Edition. W. Funk, V. Dammann, G. Donnevert
Copyright © 2007 WILEY-VCH Verlag GmbH & Co. KGaA, Weinheim
ISBN: 978-3-527-31114-9

A2.2
F-Table (95%)

f_2 \ f_1	1	2	3	4	5	6	7	8	9	10	12	15	20	24	30	40	60	120	∞
1	161.4	199.5	215.7	224.6	230.2	234.0	236.8	238.9	240.5	241.9	243.9	245.9	248.0	249.1	250.1	251.1	252.2	253.3	254.3
2	18.51	19.00	19.16	19.25	19.30	19.33	19.35	19.37	19.38	19.40	19.41	19.43	19.45	19.45	19.46	19.47	19.48	19.49	19.50
3	10.13	9.55	9.28	9.12	9.01	8.94	8.89	8.85	8.81	8.79	8.74	8.70	8.66	8.64	8.62	8.59	8.57	8.55	8.53
4	7.71	6.94	6.59	6.39	6.26	6.16	6.09	6.04	6.00	5.96	5.91	5.86	5.80	5.77	5.75	5.72	5.69	5.66	5.63
5	6.61	5.79	5.41	5.19	5.05	4.95	4.88	4.82	4.77	4.74	4.68	4.62	4.56	4.53	4.50	4.46	4.43	4.40	4.36
6	5.99	5.14	4.76	4.53	4.39	4.28	4.21	4.15	4.10	4.06	4.00	3.94	3.87	3.84	3.81	3.77	3.74	3.70	3.67
7	5.59	4.74	4.35	4.12	3.97	3.87	3.79	3.73	3.68	3.64	3.57	3.51	3.44	3.41	3.38	3.34	3.30	3.27	3.23
8	5.32	4.46	4.07	3.84	3.69	3.58	3.50	3.44	3.39	3.35	3.28	3.22	3.15	3.12	3.08	3.04	3.01	2.97	2.93
9	5.12	4.26	3.86	3.63	3.48	3.37	3.29	3.23	3.18	3.14	3.07	3.01	2.94	2.90	2.86	2.83	2.79	2.75	2.71
10	4.96	4.10	3.71	3.48	3.33	3.22	3.14	3.07	3.02	2.98	2.91	2.85	2.77	2.74	2.70	2.66	2.62	2.58	2.54
11	4.84	3.98	3.59	3.36	3.20	3.09	3.01	2.95	2.90	2.85	2.79	2.72	2.65	2.61	2.57	2.53	2.49	2.45	2.40
12	4.75	3.89	3.49	3.26	3.11	3.00	2.91	2.85	2.80	2.75	2.69	2.62	2.54	2.51	2.47	2.43	2.38	2.34	2.30
13	4.67	3.81	3.41	3.18	3.03	2.92	2.83	2.77	2.71	2.67	2.60	2.53	2.46	2.42	2.38	2.34	2.30	2.25	2.21
14	4.60	3.74	3.34	3.11	2.96	2.85	2.76	2.70	2.65	2.60	2.53	2.46	2.39	2.35	2.31	2.27	2.22	2.18	2.13
15	4.54	3.68	3.29	3.06	2.90	2.79	2.71	2.64	2.59	2.54	2.48	2.40	2.33	2.29	2.25	2.20	2.16	2.11	2.07
16	4.49	3.63	3.24	3.01	2.85	2.74	2.66	2.59	2.54	2.49	2.42	2.35	2.28	2.24	2.19	2.15	2.11	2.06	2.01
17	4.45	3.59	3.20	2.96	2.81	2.70	2.61	2.55	2.49	2.45	2.38	2.31	2.23	2.19	2.15	2.10	2.06	2.01	1.96
18	4.41	3.55	3.16	2.93	2.77	2.66	2.58	2.51	2.46	2.41	2.34	2.27	2.19	2.15	2.11	2.06	2.02	1.97	1.92
19	4.38	3.52	3.13	2.90	2.74	2.63	2.54	2.48	2.42	2.38	2.31	2.23	2.16	2.11	2.07	2.03	1.98	1.93	1.88
20	4.35	3.49	3.10	2.87	2.71	2.60	2.51	2.45	2.39	2.35	2.28	2.20	2.12	2.08	2.04	1.99	1.95	1.90	1.84
21	4.32	3.47	3.07	2.84	2.68	2.57	2.49	2.42	2.37	2.32	2.25	2.18	2.10	2.05	2.01	1.96	1.92	1.87	1.81
22	4.30	3.44	3.05	2.82	2.66	2.55	2.46	2.40	2.34	2.30	2.23	2.15	2.07	2.03	1.98	1.94	1.89	1.84	1.78
23	4.28	3.42	3.03	2.80	2.64	2.53	2.44	2.37	2.32	2.27	2.20	2.13	2.05	2.01	1.96	1.91	1.86	1.81	1.76
24	4.26	3.40	3.01	2.78	2.62	2.51	2.42	2.36	2.30	2.25	2.18	2.11	2.03	1.98	1.94	1.89	1.84	1.79	1.73

f_2 \ f_1	1	2	3	4	5	6	7	8	9	10	12	15	20	24	30	40	60	120	∞
25	4.24	3.39	2.99	2.76	2.60	2.49	2.40	2.34	2.28	2.24	2.16	2.09	2.01	1.96	1.92	1.87	1.82	1.77	1.71
26	4.23	3.37	2.98	2.74	2.59	2.47	2.39	2.32	2.27	2.22	2.15	2.07	1.99	1.95	1.90	1.85	1.80	1.75	1.69
27	4.21	3.35	2.96	2.73	2.57	2.46	2.37	2.31	2.25	2.20	2.13	2.06	1.97	1.93	1.88	1.84	1.79	1.73	1.67
28	4.20	3.34	2.95	2.71	2.56	2.45	2.36	2.29	2.24	2.19	2.12	2.04	1.96	1.91	1.87	1.82	1.77	1.71	1.65
29	4.18	3.33	2.93	2.70	2.55	2.43	2.35	2.28	2.22	2.18	2.10	2.03	1.94	1.90	1.85	1.81	1.75	1.70	1.64
30	4.17	3.32	2.92	2.69	2.53	2.42	2.33	2.27	2.21	2.16	2.09	2.01	1.93	1.89	1.84	1.79	1.74	1.68	1.62
40	4.08	3.23	2.84	2.61	2.45	2.34	2.25	2.18	2.12	2.08	2.00	1.92	1.84	1.79	1.74	1.69	1.64	1.58	1.51
60	4.00	3.15	2.76	2.53	2.37	2.25	2.17	2.10	2.04	1.99	1.92	1.84	1.75	1.70	1.65	1.59	1.53	1.47	1.39
120	3.92	3.07	2.68	2.45	2.29	2.17	2.09	2.02	1.96	1.91	1.83	1.75	1.66	1.61	1.55	1.50	1.43	1.35	1.25
∞	3.84	3.00	2.60	2.37	2.21	2.10	2.01	1.94	1.88	1.83	1.75	1.67	1.57	1.52	1.46	1.39	1.32	1.22	1.00

A2.2
F-Table (99%)

f_2 \ f_1	1	2	3	4	5	6	7	8	9	10	12	15	20	24	30	40	60	120	∞
1	4052	5000.5	5403	5625	5764	5859	5928	5982	6022	6056	6106	6157	6209	6235	6261	6287	6313	6339	6366
2	98.50	99.00	99.17	99.25	99.30	99.33	99.36	99.37	99.39	99.40	99.42	99.43	99.45	99.46	99.47	99.47	99.48	99.49	99.50
3	34.12	30.82	29.46	28.71	28.24	27.91	27.67	27.49	27.35	27.23	27.05	26.87	26.69	26.60	26.50	26.41	26.32	26.22	26.13
4	21.20	18.00	16.69	15.98	15.52	15.21	14.98	14.80	14.66	14.55	14.37	14.20	14.02	13.93	13.84	13.75	13.65	13.56	13.46
5	16.26	13.27	12.06	11.39	10.97	10.67	10.46	10.29	10.16	10.05	9.89	9.72	9.55	9.47	9.38	9.29	9.20	9.11	9.02
6	13.75	10.92	9.78	9.15	8.75	8.47	8.26	8.10	7.98	7.87	7.72	7.56	7.40	7.31	7.23	7.14	7.06	6.97	6.88
7	12.25	9.55	8.45	7.85	7.46	7.19	6.99	6.84	6.72	6.62	6.47	6.31	6.16	6.07	5.99	5.91	5.82	5.74	5.65
8	11.26	8.65	7.59	7.01	6.63	6.37	6.18	6.03	5.91	5.81	5.67	5.52	5.36	5.28	5.20	5.12	5.03	4.95	4.86
9	10.56	8.02	6.99	6.42	6.06	5.80	5.61	5.47	5.35	5.26	5.11	4.96	4.81	4.73	4.65	4.57	4.48	4.40	4.31
10	10.04	7.56	6.55	5.99	5.64	5.39	5.20	5.06	4.94	4.85	4.71	4.56	4.41	4.33	4.25	4.17	4.08	4.00	3.91
11	9.65	7.21	6.22	5.67	5.32	5.07	4.89	4.74	4.63	4.54	4.40	4.25	4.10	4.02	3.94	3.86	3.78	3.69	3.60
12	9.33	6.93	5.95	5.41	5.06	4.82	4.64	4.50	4.39	4.30	4.16	4.01	3.86	3.78	3.70	3.62	3.54	3.45	3.36

F-Table (99%) (continued)

f_2 \ f_1	1	2	3	4	5	6	7	8	9	10	12	15	20	24	30	40	60	120	∞
13	9.07	6.70	5.74	5.21	4.86	4.62	4.44	4.30	4.19	4.10	3.96	3.82	3.66	3.59	3.51	3.43	3.34	3.25	3.17
14	8.86	6.51	5.56	5.04	4.69	4.46	4.28	4.14	4.03	3.94	3.80	3.66	3.51	3.43	3.35	3.27	3.18	3.09	3.00
15	8.68	6.36	5.42	4.89	4.56	4.32	4.14	4.00	3.89	3.80	3.67	3.52	3.37	3.29	3.21	3.13	3.05	2.96	2.87
16	8.53	6.23	5.29	4.77	4.44	4.20	4.03	3.89	3.78	3.69	3.55	3.41	3.26	3.18	3.10	3.02	2.93	2.84	2.75
17	8.40	6.11	5.18	4.67	4.34	4.10	3.93	3.79	3.68	3.59	3.46	3.31	3.16	3.08	3.00	2.92	2.83	2.75	2.65
18	8.29	6.01	5.09	4.58	4.25	4.01	3.84	3.71	3.60	3.51	3.37	3.23	3.08	3.00	2.92	2.84	2.75	2.66	2.57
19	8.18	5.93	5.01	4.50	4.17	3.94	3.77	3.63	3.52	3.43	3.30	3.15	3.00	2.92	2.84	2.76	2.67	2.58	2.49
20	8.10	5.85	4.94	4.43	4.10	3.87	3.70	3.56	3.46	3.37	3.23	3.09	2.94	2.86	2.78	2.69	2.61	2.52	2.42
21	8.02	5.78	4.87	4.37	4.04	3.81	3.64	3.51	3.40	3.31	3.17	3.03	2.88	2.80	2.72	2.64	2.55	2.46	2.36
22	7.95	5.72	4.82	4.31	3.99	3.76	3.59	3.45	3.35	3.26	3.12	2.98	2.83	2.75	2.67	2.58	2.50	2.40	2.31
23	7.88	5.66	4.76	4.26	3.94	3.71	3.54	3.41	3.30	3.21	3.07	2.93	2.78	2.70	2.62	2.54	2.45	2.35	2.26
24	7.82	5.61	4.72	4.22	3.90	3.67	3.50	3.36	3.26	3.17	3.03	2.89	2.74	2.66	2.58	2.49	2.40	2.31	2.21
25	7.77	5.57	4.68	4.18	3.85	3.63	3.46	3.32	3.22	3.13	2.99	2.85	2.70	2.62	2.54	2.45	2.36	2.27	2.17
26	7.72	5.53	4.64	4.14	3.82	3.59	3.42	3.29	3.18	3.09	2.96	2.81	2.66	2.58	2.50	2.42	2.33	2.23	2.13
27	7.68	5.49	4.60	4.11	3.78	3.56	3.39	3.26	3.15	3.06	2.93	2.78	2.63	2.55	2.47	2.38	2.29	2.20	2.10
28	7.64	5.45	4.57	4.07	3.75	3.53	3.36	3.23	3.12	3.03	2.90	2.75	2.60	2.52	2.44	2.35	2.26	2.17	2.06
29	7.60	5.42	4.54	4.04	3.73	3.50	3.33	3.20	3.09	3.00	2.87	2.73	2.57	2.49	2.41	2.33	2.23	2.14	2.03
30	7.56	5.39	4.51	4.02	3.70	3.47	3.30	3.17	3.07	2.98	2.84	2.70	2.55	2.47	2.39	2.30	2.21	2.11	2.01
40	7.31	5.18	4.31	3.83	3.51	3.29	3.12	2.99	2.89	2.80	2.66	2.52	2.37	2.29	2.20	2.11	2.02	1.92	1.80
60	7.08	4.98	4.13	3.65	3.34	3.12	2.95	2.82	2.72	2.63	2.50	2.35	2.20	2.12	2.03	1.94	1.84	1.73	1.60
120	6.85	4.79	3.95	3.48	3.17	2.96	2.79	2.66	2.56	2.47	2.34	2.19	2.03	1.95	1.86	1.76	1.66	1.53	1.38
∞	6.63	4.61	3.78	3.32	3.02	2.80	2.64	2.51	2.41	2.32	2.18	2.04	1.88	1.79	1.70	1.59	1.47	1.32	1.00

A2.3
Grubbs Table

$P_{single-sided}$	90%	95%	99%
N			
3	1.148	1.153	1.155
4	1.425	1.463	1.492
5	1.602	1.672	1.749
6	1.729	1.822	1.944
7	1.828	1.938	2.097
8	1.909	2.032	2.221
9	1.977	2.110	2.323
10	2.036	2.176	2.410
11	2.088	2.234	2.485
12	2.134	2.285	2.550
13	2.175	2.331	2.607
14	2.213	2.371	2.659
15	2.247	2.409	2.705
16	2.279	2.443	2.747
17	2.309	2.475	2.785
18	2.335	2.504	2.821
19	2.361	2.532	2.854
20	2.385	2.557	2.884
21	2.408	2.580	2.912
22	2.429	2.603	2.939
23	2.448	2.624	2.963
24	2.467	2.644	2.987
25	2.486	2.663	3.009
26	2.502	2.681	3.029
27	2.519	2.698	3.049
28	2.534	2.714	3.068
29	2.549	2.730	3.085
30	2.563	2.745	3.103
31	2.579	2.760	3.119
32	2.592	2.773	3.135
33	2.605	2.787	3.150
34	2.618	2.799	3.164
35	2.630	2.812	3.178
36	2.641	2.824	3.191
37	2.652	2.835	3.204
38	2.663	2.846	3.216
39	2.674	2.857	3.228
40	2.684	2.868	3.239
$P_{double-sided}$	80%	90%	98%

A2.4
χ^2-Table

5%, 1%, and 0.1% boundaries of the χ^2-distribution:

f	5%	1%	0.1%	f	5%	1%	0.1%	f	5%	1%	0.1%
1	3.84	6.63	10.83	51	68.67	77.39	87.97	101	125.46	136.97	150.67
2	5.99	9.21	13.82	52	69.83	78.61	89.27	102	126.57	138.13	151.88
3	7.81	11.34	16.27	53	70.99	79.84	90.57	103	127.69	139.30	153.10
4	9.49	13.28	18.47	54	72.15	81.07	91.87	104	128.80	140.46	154.31
5	11.07	15.09	20.52	55	73.31	82.29	93.17	105	129.92	141.62	155.53
6	12.59	16.81	22.46	56	74.47	83.51	94.46	106	131.03	142.78	156.74
7	14.07	18.48	24.32	57	75.62	84.73	95.75	107	132.15	143.94	157.95
8	15.51	20.09	26.13	58	76.78	85.95	97.04	108	133.26	145.10	159.16
9	16.92	21.67	27.88	59	77.93	87.16	98.32	109	134.37	146.26	160.37
10	18.31	23.21	29.59	60	79.08	88.38	99.61	110	135.48	147.41	161.58
11	19.68	24.73	31.26	61	80.23	89.59	100.89	111	136.59	148.57	162.79
12	21.03	26.22	32.91	62	81.38	90.80	102.17	112	137.70	149.73	163.99
13	22.36	27.69	34.53	63	82.53	92.01	103.44	113	138.81	150.88	165.20
14	23.68	29.14	36.12	64	83.68	93.22	104.72	114	139.92	152.04	166.41
15	25.00	30.58	37.70	65	84.82	94.42	105.99	115	141.03	153.19	167.61
16	26.30	32.00	39.25	66	85.97	95.62	107.26	116	142.14	154.34	168.81
17	27.59	33.41	40.79	67	87.11	96.83	108.52	117	143.25	155.50	170.01
18	28.87	34.81	42.31	68	88.25	98.03	109.79	118	144.35	156.65	171.22
19	30.14	36.19	43.82	69	89.39	99.23	111.05	119	145.46	157.80	172.42
20	31.41	37.57	45.31	70	90.53	100.42	112.32	120	146.57	158.95	173.62
21	32.67	38.93	46.80	71	91.67	101.62	113.58	121	147.67	160.10	174.82
22	33.92	40.29	48.27	72	92.81	102.82	114.83	122	148.78	161.25	176.01
23	35.17	41.64	49.73	73	93.95	104.01	116.09	123	149.89	162.40	177.21
24	36.42	42.98	51.18	74	95.08	105.20	117.35	124	150.99	163.55	178.41
25	37.65	44.31	52.62	75	96.22	106.39	118.60	125	152.09	164.69	179.60
26	38.89	45.64	54.05	76	97.35	107.58	119.85	126	153.20	165.84	180.80
27	40.11	46.96	55.48	77	98.49	108.77	121.10	127	154.30	166.99	181.99
28	41.34	48.28	56.89	78	99.62	109.96	122.35	128	155.41	168.13	183.19
29	42.56	49.59	58.30	79	100.75	111.14	123.59	129	156.51	169.28	184.38
30	43.77	50.89	59.70	80	101.88	112.33	124.84	130	157.61	170.42	185.57
31	44.99	52.19	61.10	81	103.01	113.51	126.08	131	158.71	171.57	186.76
32	46.19	53.48	62.49	82	104.14	114.69	127.32	132	159.81	172.71	187.95
33	47.40	54.77	63.87	83	105.27	115.88	128.56	133	160.92	173.85	189.14
34	48.60	56.06	65.25	84	106.40	117.06	129.80	134	162.02	175.00	190.33
35	49.80	57.34	66.62	85	107.52	118.23	131.04	135	163.12	176.14	191.52
36	51.00	58.62	67.98	86	108.65	119.41	132.28	136	164.22	177.28	192.71
37	52.19	59.89	69.34	87	109.77	120.59	133.51	137	165.32	178.42	193.89
38	53.38	61.16	70.70	88	110.90	121.77	134.74	138	166.42	179.56	195.08
39	54.57	62.43	72.05	89	112.02	122.94	135.98	139	167.52	180.70	196.27
40	55.76	63.69	73.40	90	113.15	124.12	137.21	140	168.61	181.84	197.45
41	56.94	64.95	74.74	91	114.27	125.29	138.44	141	169.71	182.98	198.63
42	58.12	66.21	76.08	92	115.39	126.46	139.67	142	170.81	184.12	199.82
43	59.30	67.46	77.42	93	116.51	127.63	140.89	143	171.91	185.25	201.00
44	60.48	68.71	78.75	94	117.63	128.80	142.12	144	173.00	186.39	202.18
45	61.66	69.96	80.08	95	118.75	129.97	143.34	145	174.10	187.53	203.36
46	62.83	71.20	81.40	96	119.87	131.14	144.57	146	175.20	188.67	204.55
47	64.00	72.44	82.72	97	120.99	132.31	145.79	147	176.29	189.80	205.73
48	65.17	73.68	84.04	98	122.11	133.47	147.01	148	177.39	190.94	206.91
49	66.34	74.92	85.35	99	123.23	134.64	148.23	149	178.49	192.07	208.09
50	67.50	76.15	86.66	100	124.34	135.81	149.45	150	179.58	193.21	209.26

f = degree of freedom

Appendix 3

A3
Contents of the CD

To aid in working with the statistical methods and to assist laboratory work, additional Excel tables in the Office 97 format have been appended to this book on a CD, the assumption being that Excel 97 or a higher version is already installed on your computer. Because some sets of processing functions require the use of macros, macros should *not* be deactivated for security reasons.

Although there are several individual Excel tables, which can be copied and used independently of one another, the first Excel sheet from the file „Example.xls" serves as a menu with links to the other tables. For this reason, all Excel tables should be located in the same directory.

A3.1
Checklists

The tables from Section 3.5 have been copied to an interactive Excel table. The sheets are not write-protected and can be filled out with your own information. Therefore, the file

CHECKLISTS.XLS

should first be copied to the internal hard drive with the help of Windows Explorer. Double-clicking on the copied file will open it.

A3.2
Instructions for Using the Calculation Examples

The Excel workbook

EXAMPLE.XLS

offers the possibility of readily applying the given statistical procedures to your own laboratory data. The file is only for „trying out", however, and is not intended for routine laboratory data processing. The file is not based on a databank system and is therefore also unsuitable for saving procedural characteristics, measurement results, or quality assurance data. New entries will overwrite the old values.

No liability will be undertaken for the results from the Example.xls workbook. Because no rounding of intermediate results occurs in Excel, your final results may be slightly different from those values given in the book.

A3.3
Statistical Table Values

The Excel workbook

STATISTICAL TABLE VALUES.XLS

supplies frequently used statistical table values as a function of the degrees of freedom and level of significance, e.g., t-values for the calculation of confidence intervals or comparison values for the statistical hypothesis tests presented in the book.

Subject Index

a

AAS-determination, interlaboratory test 192
accreditation 6, 170, 196
– agency 5
accuracy 1 f., 202
accuracy of the mean 203
– control 146
– required 59
actual value 201
adjustment 119, 201
– cost 61
– daily 51
– instrument 51
administrative work, cost 62
aging 49
amplification range 26
analyses
– infrequent 150
– time-consuming 150
analysis effort, R-chart 96
analysis function 15
analytical method 197
analytical principle 197 f.
analytical range 16
analytical result, calculation 22
analytical sample 205
AOX 166
applicability 55
AQA 171
arbitration method 198
ARL 73
– curves 76
– cusum chart 106
– V-mask 107
assessment, analytical 169
assigned value 175
audit
– complaint-oriented 170
– documentation-oriented 170
– external 170
– internal 169
– process oriented 170
– system-oriented 170
audit report, external 167
audits 169
average run length 73
axis intercept, linear curve 18

b

batch 121
BCR 68
bias 204
biochemistry 166
BIPM 5
blank, routine monitoring 146
blank control chart 86
– decision criteria 87
blank sample 67, 146
– preliminary period 79
blank value 29
– causes 146
– determination 30
bulk sample 205

c

calibration 6, 201
– cost 61
– fundamental 15
– recalibration 9, 122
calibration certificate 119
calibration function 15, 60
– second order 19
calibration service 5
capability of detection 200
carbon balance 165
certification 6, 196
– audit 170
certified reference material 68
certified standards 120

Quality Assurance in Analytical Chemistry: Applications in Environmental, Food, and Materials Analysis, Biotechnology, and Medical Engineering, Second Edition. W. Funk, V. Dammann, G. Donnevert
Copyright © 2007 WILEY-VCH Verlag GmbH & Co. KGaA, Weinheim
ISBN: 978-3-527-31114-9

Subject Index

certifying bodies   68
chart, mean-R combination   95
client   170
climate   118
COD   80, 166
– control chart   148
COD/BOD5 relationship   166
collaborative trial   198
comparability   115
comparable conditions   203
comparable laboratory procedures   3
comparison procedure   123
complaints, handling of   174
conductivity, electrical   166
confidence interval of result   21
confidence interval, standard addition   153
confidence level   206
confidentiality, interlaboratory test   173
construction, laboratory   166
control chart   65
– blank   147
– principle   72
– recovery rate   41, 148
– revision   150
– standard deviation   148
– summary   112
– system   70
– types   80
control limits   65, 72
– Shewhart chart   83
control material
– for precision control   69
– producer   69
control samples   49, 66, 122
– applicability   70
– commercially available, requirements   69
– difference chart   97
– R-chart   92
– recovery rate chart   88
– requirements   66
– target value chart   99
– types   67
conventional true value   201
cooling chain   118
corrective actions   117
corrective measures   154
correlation, range and mean   93
cost per analysis   61
Cr, total/Cr (VI)   166
critical value   32
CRM   205

customer, complaints   117
cusum axis, scale   102
cusum chart   80
– advantages   111
– decision criteria   103
– decision criteria, numerical   108
– disadvantages   111
– establishment   101
– principle   100
– purpose   100
– V-mask   104
cuvette   120

d

decision limit   14, 30, 32, 200
decision making   2
– yes/no   4
detection limit   2
deviation   131, 203
– additive   37, 204
– multiplicative   38, 204
– random   132
– systematic   128, 132
difference chart   96
– control limits   97
– control sample   97
difference method   129
differences, successive   51
dimensional analysis   17
disposal of reagents   55
DOC   50, 166
documentation   166
drift   96
drinking water
– accuracy requirements   65
– sulfate content   66

e

equipment, cost   62
equivalency   122, 202
– different matrices   126
equivalency test
– natural samples   124
– single matrix   123
error   203
– categories   154
– constant systematic   37, 41, 204
– gross   155, 161
– proportional systematic   38, 204
– random   155, 202, 204
– systematic   1, 37, 155, 202, 204
– systematic, elimination   162
– total   203
– sources   155

Subject Index | 273

α-error   29, 32
β-error   29
EURACHEM guide   131
– concept   135
– terms   131
EUROLAB, concept   140
exclusion limit, target value chart   99
expected value   201
extraction, optimization   43
EXTRELUT   43, 45

f

F-test   207
– equivalence   123
– for linearity   24
– for outlier in calibration   27
– homogeneity of variances   25
– recovery function   40
– series   50
– training phase   63
– Youden   185
false alarm   73
Fe, total/Fe (II)   166
field methods   3
field test, result   164
filter   120
forensic medicine   2
frequency of usage   62
function coefficients, second order curve   20

g

Gas chromatograph   120
gas chromatography   46
Gaussian curve   74, 82
GLP   4
GMP   170
Good Laboratory Practice   4, 170
Good Manufacturing Practice   170
gross error   204
Grubbs outlier test, equivalency   127
Grubbs test   171, 208
– difference method   129
GUM   131

h

halogenated hydrocarbons   166
HAMPEL estimator   189
homogeneity
– of sample   16
– of variances   25
HPTLC   43, 45
ICP-AES   46

i

identification   145
impartiality, interlaboratory test   173
imprecision
– maximum relativ   35
– relative analytical   29
influence
– matrix   14
– time   14
influencing variables   131
inspection   117
installation   166
– quality   117
instrument, calibration   117
intercept, recovery function   39
interference   38
– matrix   37
interfering substances   60
interlaboratory test   6, 136, 170, 198
– evaluation   176
– ISO 5725-2   177
– ISO guide 43   188
– planning   174
– planning and execution   173
– process standardization   171
– provider   173
– samples   175
– use of results   178
– validation   178
interlaboratory trial   169, 198
internal standard, procedure   47
internal standards   46
– requirements   47
ion balance   165
IR-spectrophotometer   120
ISO/TS 21748, estimation of uncertainty   137

k

k-Faktor   35

l

laboratory monitoring, Youden   179
laboratory
– comparison   65
– reference   170
limit of determination   200
limit of precision, interlaboratory test   177
limit of quantification   14, 16, 30, 35, 61, 200
– approximation   36
– calculation   36
limit value, monitoring   143

Subject Index

linearity 14
– testing 14
– verification 23
log-book 167
lower range limit 27

m

maintenance 117, 167
– cost 62
– preventive 120
management
– laboratory 116, 118
– responsibility 116
Mandel's fitting test 23
material, cost 62
matrix 16, 205
– effects 151
– influences 37
mean control chart 80, 85
mean, subgroup 74
measurement principles 197
measurement sample 206
measurement standard 200
measurement uncertainty 16, 130, 204
measuring accuracy, Youden 185
measuring device 200
measuring equipment 119
measuring instrument 26
microburette 64
minimal detectable value 200
minimum detectable value 14, 16, 34, 61
multiple curve fitting 46
multiple regression 46

n

National Accreditation Council 5
natural sample 67
– preliminary period 79
– spiked 79
Neumann, trend test 51
NIST 68
nitrite, example 21
nitrogen balance 165
nomogram, cusum chart, decision criteria 110
non-linearity 16
NORDTEST, concept 138

o

OECD 4, 170
one-tailed distribution 33
orientation test 3, 4

out of control situation
– cusum chart 103
– Shewhart chart 84
outlier
– interlaboratory test 171
– test 208
– types 171
outlier test
– calibration 26
– Youden 186
oxidation state 16

p

personnel 119
– training 119, 167
– training, cost 62
pH meter 120
phase I 9
– results 14
– summary 54
phase II 57
phase III 115
phase IV 169
4-phase model 6
phenobarbital 43
phosphorous balance 165
physiology 166
pipette 119
– disposable 64
– volumetric 64
pipetting step, optimization 64
plausibility 155
plausibility check 164
– consequences 166
precision 1, 14, 202
– between batch 203
– from lab to lab 203
– serial 202
– testing 14
– under comparable conditions 203
– within a series 203
– within-batch 202
precision control, R-chart 149
precision requirements 4
prediction interval 27
preliminary period 72, 78
primary standard 200
probability
– of error 29
– outside control limits 74
probability of error 206
procedure
– analytical 59
– ready-to-use 3

process data   9
– characteristic   171
process standard deviation, second order curve   21
process variation coefficient   19
proficiency test   171, 188
proficiency testing, evaluation   175
prognosis interval   22
project report   167
proportional systematic error, matrix-dependent   87
purpose, analytical   2

q
quality   195
quality assessment, external   196
quality assurance   195
– documentation   167
– external   5, 169
– routine   115
quality characteristic   195
quality control   117, 195
– internal   196
– routine   145
– statistical   196
quality improvement   195
quality management   195
quality management system   196
– interlaboratory test   173
quality manual   117
quality objectives   64
quality planning   195
quality requirements
– external   65
– internal   66
question, analytical   4

r
R-chart   80, 89, 148
– decision criteria   91
– establishment   90
range control chart   89
range
– analytical   14, 16, 60
– of interest   59
readjustment   55
reagents   117
recalibration   55
recovery control chart   87
– construction   89
recovery curve   39
recovery function   37
– application   41
– establishment   38

– example   43
– matrix effects   45
– prerequisites   39
recovery rate   41, 199
– control chart   148
– mean   89
– preliminary period   79
recovery rate chart
– control limits   89
– control samples   88
– warning limits   89
reference laboratory   5
reference material   148, 205
– certified   205
reference method   198
reference procedure   123, 198
reference value, cusum chart   101
Refractometer   120
regression analysis, linear   18
regression model   23
regression, orthogonal   126
reliability   115, 196
repeatability   202
repeatability standard deviation   177
repeatable conditions   173, 203
report of analytical results   145
representativeness of sample   16
reproducibility   203
reproducibility standard deviation   177
reproducible conditions   173
residual analysis   24
residual standard deviation   19, 24
– second order curve   20
resolution   199
result
– calculation   21
– calculation 2nd order curve   22
robust evaluation procedures   172
robustness   199
round robin test   65, 170, 198
routine   115
ruggedness   199

s
s-chart
– central line   98
– control limits   98
s-chart, see standard deviation chart   98
S/N ratio   37
safety, occupational   55
sample
– data   164
– digestion   37

– extraction   37
– interlaboratory test   175
– modification   38
– path   118
– preservation   164
– qualification   121
– standard   16
– storage   17
– for R-chart   149
– for Youden evaluation   180
scales   119
scatter *see* random error   155
screening   2
screening procedures   3
secondary standard   200
selectivity   1, 60
– analytical   199
selenium, determination   45
sensitivity   1, 18, 60
– analytical   199
– second order curve   20
serum, human   45
Shewhart   70
Shewhart chart   80-81
– application   85
– construction   82
– statistical fundamentals   81
SI system   5
significance level   34, 206
single-sided question   33
slope   60
– dimension   17
– linear curve   18
– recovery function   39
solvent   121
SOP   117, 167
specificity   1 f., 60
– analytical   199
specimen   205
spectrofluorometer   120
spectrophotometer   120
spiked natural sample   88
spiking   45
– interlaboratory test   174
spiking, *see* standard addition   151
– quantity   87
– substance   88
– volume   152
stability   17
– control sample   66
staff
– change   9
– experience   150
standard   200

standard addition   41
– calculation   152
– method   3, 198, 151
– multivariate   46
standard deviation
– assigned   189
– between batch   49
– between laboratories, Youden   185
– chart   80
– laboratory internal, Youden   185
– procedures   122
– process   60
– residual   60
– sample   16
– total   49
– within batch   49
standard deviation chart, *see* s-chart   98
standard solution   67
– preliminary period   79
standardization, process   171
storage   118
successive dilution   17
summation sign   18
surface water, information   164
synthetic sample   68
– interlaboratory test   174
systematic error   61
– Youden plot   180

t

t-factor   33
t-test   206
– coefficients of recovery function   40
– equivalency   126
– for outlier in calibration   27
– joint   129-130
– means   207
target value   201
target value chart   80, 99
– control sample   99
test, statistical   206
test report, uncertainty   142
testing value x_a, standard addition   153
thermometer   119
threshold limit   4
time dependency   49
time-consuming analyses   150
total standard deviation, Youden   186
toxicology   2
trace analysis   118
traceability   117, 200
training phase   62
trend test   51

trouble-shooting   155
– scheme   162
true value   201
trueness   1, 202 f.

u

uncertainty
– combined   132, 204
– expanded   132
– maximally allowed   35
– measurement   204
– of measurements   130
– prelaboratory steps   142
– standard   132, 204

v

V-mask   104
– parabolic   108
– parameters   105
validation   117, 196
VAM, concept   139
variance   14
– inhomogeneity   16
– homogeneity of   16
variance analysis, preliminary period   80
verification   196

w

warning limit, revision   150
warning limits   65, 72
– Shewhart chart   83

waste disposal   117, 167
waste water   166
– information   164
water, purity   121
weighted regression   46
working range
– preliminary   16

x

x-chart   148

y

Youden
– diagram   182
– Ellipse   181
– example   187
Youden diagram, evaluation   182
Youden evaluation
– advantages   186
– disadvantages   186
Youden method, interlaboratory test   179
Youden plot   179
– normalized evaluation   181
– simplified   183
– three and four dimensional   184

z

Z score   190
– interpretation   190

Related Titles

J. Ermer, J. H. McB. Miller (Eds.)

Method Validation in Pharmaceutical Analysis
A Guide to Best Practice

2005
ISBN 978-3-527-31255-9

T. A. Ratliff

The Laboratory Quality Assurance System
A Manual of Quality Procedures and Forms

2003
ISBN 978-0-471-26918-2